Practical Channel-Aware Resource Allocation

Michael Ghorbanzadeh • Ahmed Abdelhadi

Practical Channel-Aware Resource Allocation

With MATLAB and Python Code

 Springer

Michael Ghorbanzadeh
Department of Commerce
National Telecommunications and
Information Administration
Washington, DC, USA

Ahmed Abdelhadi
Electrical and Computer Engineering
University of Houston
Houston, TX, USA

ISBN 978-3-030-73634-7   ISBN 978-3-030-73632-3   (eBook)
https://doi.org/10.1007/978-3-030-73632-3

This Springer imprint is published by the registered company Springer Nature Switzerland AG
The registered company address is: Gewerbestrasse 11, 6330 Cham, Switzerland

To my wife Diane for all her sacrifices and unconditional love.

Michael

Preface

This volume addresses the concept of radio resource allocation for cellular communications systems operating in congested and contested environments with an emphasis on end-to-end quality of service (QoS). The radio resource allocation is cast under a proportional fairness formulation which translates to a convex optimization problem. Moreover, the resource allocation scheme considers subscription-based and traffic differentiation in order to meet the QoS requirements of the applications running on the user equipments (UEs) in the system. The devised resource allocation scheme is realized through a centralized and distributed architecture, and solution algorithms for the aforementioned architecture are derived and implemented in the mobile devices and the base stations. The sensitivity of the resource allocation scheme to the temporal dynamics of the quantity of the users in the system is investigated. Furthermore, the sensitivity of the resource allocation scheme to the temporal dynamics in the application usage percentages is accounted for. In addition, a transmission overhead of the centralized and distributed architectures for the resource allocation schemes is performed. Furthermore, the resource allocation scheme is modified to account for a possible additive bandwidth done through spectrum sharing in congested and contested environments, in particular spectrally coexistent radar systems. The radar-spectrum additive portion is devised in a way to ensure fairness of the allocation, high bandwidth utilization, and interference avoidance. In order to justify the aforesaid modification, the interference from radar systems into the long-term evolution (LTE) as the predominant 4G technology is studied to confirm the possibility of spectrum sharing. The preceding interference analysis contains a detailed simulation of radar systems, propagation path loss models, and a third-generation partnership project compliant LTE system. The propagation models are free space path loss (FSPL) and irregular terrain model (ITM). The LTE systems under consideration are macro cell, outdoor small cells, and indoor small cells. In order to consider the QoS on an end-to-end basis, a delay-based modeling of the backhaul network, as the bottleneck in modern cellular communications systems, is presented. The modeling task is fined-grained through an optimum selection from a variety of candidate hidden Markov models and vector quantization schemes. A model-based signature statistical analysis is performed to evaluate the modeling fidelity.

Washington, DC, USA   Michael Ghorbanzadeh

Houston, TX, USA   Ahmed Abdelhadi
May 2021

Introduction

The unceasing magnitude of mobile broadband users on one side and the sheer volume of the traffic which they generate on the other side has pushed the demands by mobile broadband networks to extents beyond any figures network planners ever envisaged. Such an ever-increasing trend in the mobile broadband subscriptions stems from widespread prevalence of mobile broadband smart devices whose applications consume enormous amounts of bandwidth. However, communications networks possess finite resources such as the radio frequency spectrum and transport backhaul, expensive and shared among numerous users of many services. While there has been several federal initiatives to share spectrum used by government entities with the mobile broadband services, the demand for more bandwidth is only to increase in au courant communications systems. Besides, while such capacity-additive approaches can cater to the mobile broadband traffic growth, mobility properties such as intermittent connectivity and location shifting, coupled with a proliferation of smart devices and inexpensive mobile data tariffs and compounded by the emergence and prevalence of the quadruple play prevalent in the fourth, fifth, and upcoming sixth generation, aka 4G, 5G, and 6G, cellular networks, create numerous field challenges to achieving a sublime end-to-end network performance. Hence, modern-day communications systems of mobile broadband devices ought to be far more intelligent in handling the traffic growth.

Since the voluminous mobile broadband traffic incorporates traditional multi-play services such as Internet Protocol (IP) Television, voice-over-IP, and video conferencing which subsume fairly well-known quality of service requirements, such a hybrid mobile broadband traffic must be treated with relevant running application quality of service requirements in mind. Moreover, the evolution of the modern cellular communications systems into an all-IP paradigm in 5G and 6G introduces new challenges for the traditional services such as telephony, which require a minimum level of performance to be guaranteed. Additionally, the uptake of cloud-based and web-hosted services necessitates a new quality of service requirements for the network performance. Moreover, differentiating mobile subscribers is of high consequence; some statistics indicate that less than 10% of the mobile broadcast users generate the bulk of the traffic due to the nature of the services they employ. On the other hand, the temporal usage patterns of the applications which generate the mobile broadband traffic are of prominent importance to intelligent cellular communications systems tackling the complexities of the present-day mobile Internet traffic. A case in point is the voluminous amount of signaling traffic produced by smart phones in social networking and instant messaging applications which intermittently connect to the network in the background. These situations will be further exacerbated by the advent of mobile gaming, which not only does produce an excessive interactive signaling over extended time durations, but also conspicuously escalates the volume of data traffic in the network. Some famous game servers could carry petabytes of data up to a few years ago on a daily basis and shifting these into smart devices will push the data traffic higher than ever conceived.

Inasmuch as the aforementioned discussions germane to over-provisioning in the cellular system from the bandwidth viewpoint, traffic and subscriber differential treatments to meet the quality of service requirements, proper treatment of the traffic in the backhaul, and considering wireless channel conditions are integral to elevating the quality of experience in the cellular communications ecosystem. While extending the capacity of the mobile broadband services is a must, in observation of the perpetually increasing smart device generated traffic, optimizing network efficiency with quality of service triaging to maximize the utilization of the resources is extremely desired. Introducing efficient and dynamic traffic management as resource allocation techniques not only improves the quality of experience, it further reduces the subscriber churn and proves a business opportunity for operators that make an effort in leveraging such quality of service minded technologies in their networks. The quality of service mechanisms must span from the core network to the radio access network to preclude congestion in any portion of the system. However, this effort is wildly facilitated by the emergence of virtualization and software-radio so prevalent in modern day 5G networks.

This book addresses the concept of modern radio resource allocation for cellular communications systems, to be implemented in intelligent software-based schedulers, operating in congested and contested environments with an emphasis on the quality of service. The book aims at creating a theoretical and practical framework of radio resource allocation optimization cast under a proportional fairness formulation such that subscription-based and traffic-type differentiation as well as temporal usage changes are taken into equation to fulfill the quality of service requirements of applications hosted on the user equipment, thereby elevating the user quality of experience and decreasing the subscriber churn rate. The devised novel resource allocation modus operandi is realized in centralized and distributed architectures, retains mathematical convexity, and has tractable solution algorithms implementable in the next-generation mobile devices and base stations. The sensitivity of the resource allocation technique to the temporal dynamics in the user quantity and application usage is investigated, and the transmission overhead of the aforementioned centralized and distributed architectures is studied.

Moreover, while the book provides enhancements of the formulation to radio resource block assignment to make the resource allocation approach even more pragmatic, it provides simple solution algorithms for such feature enhancements. Besides, the impact of channel conditions reflected through the radio environment map is added to the resource allocation approach such that the resource assignment procedure not only accounts for the traffic type, subscription weights, and temporal usage changes, but renders the resource assignment through the lens of the users and their channel conditions. The impact of channel is sifted by leveraging the irregular terrain model which relies on precise path elevations so as to shift the resource allocation from a theoretical realm into a field-oriented circumstance in which terrain conditions directly speak to the user channel quality according to which the resource allocation is performed.

Michael Ghorbanzadeh

Contents

1 Quality of Service and Resource Allocation in Communication Systems 1
   1.1   Introduction 1
   1.2   Current Trends in End-to-End QoS in Cellular Networks 4
   1.3   Network Quality of Service Challenges 8
   1.4   Focus of This Book 9
   1.5   Book High Level Contribution and Organization 12
   References 14

2 Utility Functions and Radio Resource Allocation 17
   2.1   Introduction 17
   2.2   Radio Resource Allocation Literature Survey 18
   2.3   Application Utility Functions 20
         2.3.1   Application Utility Function MATLAB Code 22
   2.4   Proportional Fairness and Frank Kelly Algorithm 23
   2.5   Chapter Summary 24
   References 25

3 Resource Allocation Without Channel 31
   3.1   Introduction 31
   3.2   Resource Allocation: Centralized Architecture and Solution 32
         3.2.1   Centralized Architecture Simulation 39
         3.2.2   Centralized Resource Allocation in Real-World Implementation 41
         3.2.3   MATLAB Code for Centralized Resource Allocation 45
   3.3   Resource Allocation: Distributed Architecture and Solution 48
         3.3.1   Distributed Resource Allocation Simulation 56
         3.3.2   MATLAB Code for Distributed Resource Allocation 60
   3.4   Mathematical Equivalence 62
   3.5   Distributed or Centralized 66

3.6   Benchmark Comparison   66
3.7   Chapter Summary   68
References   69

4   Radio Resource Block Allocation   71
4.1   Introduction   71
4.2   Continuous Optimization Relaxation   72
4.3   Radio Resource Block Allocation Optimization and Solution   74
        4.3.1   Radio Resource Block Allocation Simulation Results   79
        4.3.2   RB Allocation MATLAB Code   84
4.4   Chapter Summary   90
References   91

5   Resource Allocation with Channel   93
5.1   Introduction   93
5.2   Channel Quality, Modulation, and Coding   94
5.3   Radio Resource Management   97
5.4   Resource Allocation Efficacy and Channel Conditions   99
5.5   Channel-Aware Distributed Resource Allocation Formulation   101
        5.5.1   Determining ϵ   102
        5.5.2   Determining K   104
5.6   Global Solution Existence   104
        5.6.1   Solution for Channel-Aware EURA Optimization   106
        5.6.2   IURA Global Optimal Solution   109
5.7   Simulation Results   110
        5.7.1   MATLAB Code   111
5.8   Chapter Summary   120
References   121

6   Propagation Modeling   123
6.1   Introduction   123
6.2   Geodesic Calculations   124
6.3   Databases   129
        6.3.1   Terrain Profile   129
        6.3.2   Surface Refractivity and Climate   131
6.4   ITM Propagation Pathloss   135
6.5   Python Code   136
        6.5.1   Code Base Installation   137
        6.5.2   Propagation Code Run   139
6.6   Chapter Summary   140
References   140

7   Channel-Aware Resource Allocation Large Scale Simulation   143
7.1   Introduction   143
7.2   Large Scale Network Simulation   144
7.3   Chapter Summary   218
References   220

8 Book Summary 223

Index 227

Acronyms

RF	Radio frequency
TVWS	Television white space
VHF	Very low frequency
UHF	Ultra high frequency
P2P	Point-to-point
P2PM	Point-to-point mode
APM	Area prediction mode
ITM	Irregular terrain model
IP	Internet protocol
QoS	Quality of service
QoE	Quality of experience
RAN	Radio access network
VPI	Virtual path indicator
QCI	QoS class identifiers
FDMA	Frequency division multiple access
VCI	Virtual circuit indicator
WSQM	Web service quality model
BE	Best effort
AF	Assured forwarding
EF	Expedited forwarding
M2MC	Machine-to-machine communications
AV	Audio and video
MMSE	Minimum mean-squared error
OFDM	Orthogonal frequency division multiplexed
IoT	Internet-of-Things
CQI	Channel quality indicator
MCS	Modulation-coding scheme
OPEX	Operation expenditure
AP	Access point
RB	Resource broker
PC	Personal computer

VM	Virtual machine
PCI	Peripheral component interconnect
Mbps	Megabit per second
kbps	Kilobit per second
s	Second
m	Meter
S	Siemens
deg	Degrees
RB	Resource block
MEC	Multi-access edge computing
RRE	Radio resource element
AMC	Adaptive modulation and coding
RMM	Rate matching module
NF	Noise figure
DSP	Digital signal processing
SRS	Sounding reference signal
RM	Resource management
PUCCH	Physical uplink control channel
PUSCH	Physical uplink shared channel
REM	Radio environment map
SNR	Signal-to-noise ratio
GHz	Gigahertz
MHz	Megahertz
kHz	Kilohertz
DB	Database
dB	Decibel
dBm	Decibel Milli
dBi	Decibel Milli referenced to isotropic antenna
TX	Transmitter
RX	Receiver
AGL	Above ground level
AMSL	Above mean sea level
kB	Kilobyte
WGS	World geological survey
WGS84	World geological survey 1984
DEM	Digital elevation model
NED	National elevation dataset
arcs	Arc second
USGS	United states geological survey
ITU	International Telecommunications Union
ITU-R	International Telecommunications Union Radiocommunication Sector
ITU-R P	International Telecommunications Union Radiocommunication Sector Radiowave Propagarion
DOD	Department of defense

SAS	Spectrum access system
PyShp	Python shape-file library
ESRI	Environmental Systems Research Institute
JSON	JavaScript object notation
SSL	Open secure sockets layer
psutil	Process and system utilities
CPU	Central processing unit
API	Application programming interface
GB	Gigabyte
LFS	Large file storage
EIRP	Effective radiated isotropically radiated power
LCLU	Land use land cover
K	Kelvin
EB	Exabytes
MSO	Multiple system operator
CBRS	Citizen broadband radio services
UMTS	Universal mobile terrestrial system
3GPP	Third generation partnership project
MNO	Mobile network operator
6G	Sixth generation
5G	Fifth generation
4G	Fourth generation
QoS	Quality of service
QoE	Quality of experience
IPTV	Internet protocol television
VoIP	Voice-over-IP
LTE	Long-term evolution
LTE-A	LTE-advanced
WiMAX	Worldwide interoperability for microwave access
RAN	Radio access network
RSVP	Resource reservation protocol
IntServ	Integrated services
DiffServ	Differentiated services
IPv6	IP version 6
IPv4	IP version 4
DSCP	Differentiated code point
EF	Expedited forwarding
AF	Assured forwarding
UE	User equipment
BS	Base station
eNB	Evolved node-B
eNodeB	Evolved node B
MS	Mobile Device
VLL	Virtual leased line
AP	Access point

AR	Access router
PHY	Physical
MAC	Medium access control
PAPR	Peak-to-average power ratio
RRB	Radio resource block
RB	Resource block
OFDMA	Orthogonal frequency division multiple access
SC-OFDMA	Single carrier OFDMA
FFT	Fast fourier transform
IFFT	Inverse FFT
DPI	Deep packet inspection
WSQM	Web service quality model
OSI	Open system interconnection
TCP	Transport control protocol
UDP	Universal datagram protocol
RRM	Radio resource management
BER	Bit error rate
CAC	Call admission control
VCI	Virtual circuit indicator
ATM	Asynchronous transfer mode
UL	Uplink
DL	Downlink
FCC	Federal communications commission
NTIA	National Telecommunications and Information Administration
PCAST	President's Council of Advisers on Science and Technology
NPRM	Notice of proposed rulemaking
FNPRM	Final notice of proposed rulemaking
FTP	File Transfer protocol
SMTP	Simple mail transfer protocol
inf	Inflection
FCFS	First-come first-serve
WFQ	Weighted fair queuing
SINR	Signal-to-interference-plus-noise-ratio
SNR	Signal-to-noise ratio
BLER	Block error rate
VM	Virtual machine
HTTP	Hyper text transfer protocol
EURA	External UE resource allocation
IURA	Internal UE resource allocation
MME	Mobility management entity
FSPL	Free space path loss
ITM	Irregular terrain model
InH	Indoor hotspot
UMa	Urban macro
UMi	Urban micro

SMa	Suburban macro
EESM	Exponential effective SINR mapping
FDD	Frequency division duplexing
TDD	Time division duplexing
PRI	Pulse repetition interval
TTI	Transmission time interval
MIMO	Multi input multi output
HARQ	Hybrid automatic repeat request
Hz	Hertz
LoS	Line-of-sight
NLoS	Non-line-of-sight
km	Kilometers

Chapter 1
Quality of Service and Resource Allocation in Communication Systems

1.1 Introduction

Year after year, mobile broadband users' quantity and their generated data traffic volume have observed an unceasing increase. Ericsson [1] has shown global total monthly mobile data traffic approximated 33 exabytes (EB) by the end of 2019 and projected a fivefold growth by 2025 to about 164 EB per month, which illustrates the mobile data consumption of a tad 6 billion people utilizing smartphones, laptops, and a miscellaneous of au courant devices of that ilk. According to Ericsson [1], smartphones will remain the epicenter of this data generation trend since they produce 95% of the current mobile data traffic whose share is expected to increase throughout the forecast time period until 2025. Moreover, markets launching the Fifth Generation (5G) technology will potentially lead to traffic growth during the forecast by 2025, anticipating the 5G to carry 45% of mobile data traffic. Besides, Ericsson and others [1–3] believe that mobile broadband traffic growth can vary significantly among different countries based on the dynamics of their local markets. For instance, while the growth rate of US traffic declined in 2018, it rebounded to formerly forecast rates in 2019, whereas China observed a record high traffic growth in 2018 and India had an upward and consistent trajectory of data growth in the same year. Ericsson's report [1] relates the mobile data traffic growth to device capability enhancements, uptick of data-heavy traffic content/services, and inexpensive data plan availability.

Ericsson's [1] results are depicted in Fig. 1.1. Such a dramatically incremental trend in the mobile broadband traffic stems from the prevalence of mobile broadband smart devices that produce enormous traffic causes grave concerns for Mobile Network Operator (MNO) and Multiple System Operator (MSO) industries. Such concerns have spawned a perpetual demand for larger radio frequency (RF) spectrum assignment to mobile broadband services. On the other hand, RF spectrum is scarce and backhaul capacity is expensive. The RF spectrum scarcity has led to government initiatives to avail share spectrum operation on a non-interference basis

© The Author(s), under exclusive license to Springer Nature Switzerland AG 2022
M. Ghorbanzadeh, A. Abdelhadi, *Practical Channel-Aware Resource Allocation*,
https://doi.org/10.1007/978-3-030-73632-3_1

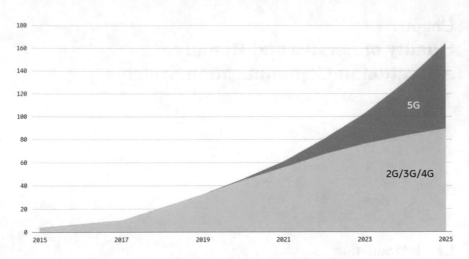

Fig. 1.1 The mobile broadband data growth by 2025 surpasses 164 EB [1]

in order to further accommodate the demands by mobile broadband services. Examples of these are the Citizen Broadband Radio Services (CBRS) as a framework to enable sharing between mobile broadband with federal ship-borne and ground-based radar and space-borne satellite systems, Television White Space (TVWS) that seeks to share RF spectrum between mobile broadband with Very Low Frequency (VHF) and Ultra High Frequency (UHF) television stations, 6 GHz band that seeks to create a sharing framework between mobile broadband and microwave point-to-point (P2P) links, and the 5.9 GHz band that tries to motivate sharing between mobile broadband with intelligent transportation public safety and radar systems.

While the need for more bandwidth is one issue, mobility requirements coupled with the proliferation of smart devices and inexpensive mobile data tariffs, compounded by the emergence and prevalence of the quadruple play, and the widespread deployment the Fourth Generation (4G) and 5G cellular communications networks have created many challenges in achieving a satisfactory user experience, good end-to-end network performance, and a profitable operator revenue [2, 4]. Hence, communication systems should be more intelligent than the traditional cellular networks in handling the traffic growth. Since mobile broadband traffic nature is of a wide variety such as traditional multi-play services like voice-over-IP (VoIP), Internet Protocol (IP) Television (IPTV), and video conferencing with well-known quality of service (QoS) requirements, it should be treated with QoS requirements of the services generating the traffic in mind. Besides, the presence of cellular network services in an all-IP paradigm of 4G and 5G introduces challenges for traditional telephony services that require a minimum level of performance. indexvoice-over-IP,VoIP

Furthermore, the widespread prevalence of the cloud services should facilitate many QoS challenges via software radio implementation of infrastructure features in a scalable robust fashion on the cloud. For instance, using cloud software implementation, we can easily differentiate mobile subscribers; Zokem [5] reports that less than 10% of the mobile broadband users subscribed to a relatively small number of monthly minutes produce the lion's share of the traffic due to type of the services they use. Such temporal usage patterns of the applications that generate mobile broadband traffic should account for complexities of the present-day mobile traffic. For instance, a large volume of smartphone signaling traffic from social networking and instant messaging applications intermittently connecting to the network in the background many times a day is a major challenge. Moreover, advent of mobile gaming applications not only producing an excessive interactive signaling over extended periods but also adding to the traffic volume is another major challenge; an example is Zynga servers carrying 0.82 petabyte (PB) of daily data [6, 7]. Shifting of this traffic completely to smart devices pushes the data traffic beyond any figures network planners ever would conceive.

A takeaway from the discussions above in the context of the mobile Internet traffic is the occurrence of severely congested Radio Access Networks (RANs) and the making of backhaul into a bottlenecks of cellular communication systems. The RAN becomes a bottleneck as the radio spectrum is scarce and the users' traffic transfers to and from the base stations in the uplink (UL) and downlink (DL) directions, respectively. This traffic includes signaling spikes killing the bandwidth as well as the application-generated traffic that has various QoS requirements and encompasses temporal dynamics based on the user's focus on the device. Furthermore, the mobile broadband traffic streams aggregate into the backhaul, whose congestion is exacerbated by the fact that it consists chiefly of legacy networks that can adversely impact the QoS of the traffic by introducing excessive delays, jitter, and packet loss. These performance parameters should be incorporated into modern cellular systems to ensure the backhaul QoS, which directly impacts the applications performance, and enhance the users' quality of experience (QoE), which is their perception of the cellular communications networks end-to-end performance and speaks directly to the subscriber churn and operator revenue.

Last but not the least, while over-provisioning bandwidth is part of a solution, differentiating traffic and users based on their QoS requirements and properly dealing with the backhaul traffic are integral to elevating the QoE for the cellular communication systems. To recap, even though extending the capacity of the network for the mobile broadband services is indispensable due to the ever-increasing smart device-generated traffic, the MNOs should optimize their network efficiency via QoS treatments, which improves the QoE in the network and efficiently maximizes the network resource utilization. Leveraging efficient dynamic traffic management and resource allocation techniques can remedy users' QoE, which reduces the subscriber churn; meanwhile, it proves a business case for operators to make an effort in leveraging QoS-related technologies in their networks. The QoS mechanisms, implemented in the RAN and the core, must span from the core network to the RAN in order to eschew congestion occurrences in any portion of the network because a localized congested section might adversely affect the overall network performance and degrade the QoE. The remainder of this chapter is organized as follows. Section 1.2 provides information on current trends for QoS in cellular communication systems. Section 1.2 conducts a literature survey on the current research in the context of the QoS in cellular communication networks [7]. Section 1.3 discusses various challenges associated with the QoS in cellular systems. Section 1.4 discusses the contributions of this book. Finally, Sect. 1.5 presents the organization and contributions of this book.

1.2   Current Trends in End-to-End QoS in Cellular Networks

In this section, we conduct a literature survey on the current research in the context of the QoS in cellular communication networks. In particular, we regard certain QoS solutions for the all-IP wired and wireless networks. The challenges that have been introduced by the widespread deployment of the smart devices in cellular communication systems (Chap. 2) have led to an upsurge in the research works that target QoS in the RAN. In addition, Long Term Evolution (LTE) [8, 9] has become the paramount 4G technology widely deployed all over the globe. Hence, many of the current studies of RAN QoS are tailored to the LTE infrastructure. Such works usually focus on a specific layer of the protocol stack including physical (PHY), link layer, network, or application layer.

Piro et al. [10] developed a framework for the QoS at the link level. Their approach used the discrete control theory to create a maximum throughput scheduler in the DL direction for the real-time multimedia services in the LTE networks. Marchese et al. [11] used a link layer QoS method to adapt the bandwidth, which is going to be allocated to a buffer, for conveying the heterogeneous traffic. The authors in [12] proposed a traffic segmentation approach to quantify the QoS at a given QoE in terms of the spectral efficiency, cost, and resource over-provisioning. In their work, the core network classified incoming packets based on a deep packet inspection (DPI) and marked the packet header using the DSCP. Then, the packets were scheduled using queue weights.

Next, Larmo et al. [13] focused on the PHY and link layers' design of the LTE. They indicated that deploying the single-carrier orthogonal frequency division multiple access (SC-OFDMA) [14] in the UL direction improves (reduces) the peak-to-average power ratio (PAPR) in the LTE system, which increases the spectral efficiency for the cell-edge users or increases the cell radius. The authors also mentioned that leveraging 12 subcarriers and 180 kHz channels in the radio resource blocks (RRBs) of the LTE generates very narrow frames. Thus, users' UL data transmissions may be scheduled depending on the channel quality at the current state.

Next, Ali et al. [15] proposed a scheduler using the game theory. The first level distributed RRBs are among the classes with different QoS requirements; then, a delay-based air interface scheduler in the second level fulfilled delay requirements defined for the LTE classes. Then, a cooperative game between various service class flows was done, and a Lagrangian formulation was used to find an associated Pareto optimality [16]. The delay-based scheduler would check each user's packet delays within its respective service class and make scheduling decisions in the DL direction based on the current channel conditions. Next, the authors in [17] suggested a mathematical framework based on the supply function bidding in microeconomics theory [18]. Their method enabled users to opportunistically compromise the efficiency based on their demands such that each user is satisfied with a lesser QoS whilst users' social welfare is maximized. Monghal et al. [11] introduced a decoupled time–frequency orthogonal frequency division multiple access (OFDMA) [19] packet scheduler for the LTE to control the throughput fairness among the users.

Some QoS works in the cellular systems rely on the network layer. For example, Lee [20] studied QoS-based routing in a multi-hop wireless network with an eye on energy efficiency. He investigated the effect that various routing methods produced from an energy efficiency perspective. In particular, he demonstrated his method accompanies trade-offs among the end-to-end delay, throughput, and energy efficiency and suggested that the shortest path routing is a good candidate once the delay, bandwidth, and energy are considered jointly. Similarly, Ekstrom [21] surveyed the QoS provisioning for traffic separation within the 3rd generation partnership project (3GPP) standardization [22], which ultimately led to those of the LTE, i.e. the bearers and service flows. Next, Alasti et al. [23] explained about the QoS provisioning built within the Worldwide Interchangeability for Microwave Access (WiMAX) [24] and LTE, as two then-competing 4G technologies. They discussed service flows and bearers as QoS mechanisms aim at the network layer.

Then, the authors in [25] studied the LTE backhaul QoS and showed that the backhaul creates a bottleneck. He suggested leveraging QoS Class Identifiers (QCI)-DSCP as well as the policing, scheduling, and shaping of the flows to provide with QoS over the backhaul. On the other hand, the authors in [26] investigated the QoS at the network layer from a policy management perspective, which emphasized on the traffic and subscriber differentiation. Moreover, they investigated the operator revenue by generating business models for the QoS and suggested adoption of the policy management approach, which firstly concentrated on reducing the congestion and then on applying the policies per subscriber and even per data flow. Next, Li [27] investigated the QoS, traffic management, and resource management in Universal Mobile Terrestrial System (UMTS) [22] and suggested a distributed infrastructure in the user equipment and base stations (BSs) to provision QoS in UMTS networks. He proposed shaping UE traffic if necessary. He also applied traffic shaping at the BSs to deal with congested cells. Then, Gorbil et al. [28] suggested a network layer QoS solution to support real-time traffic in wireless networks. They leveraged a hybrid routing protocol to enable QoS traffic support for real-time traffic via an educated path selection at source nodes.

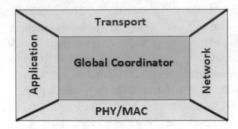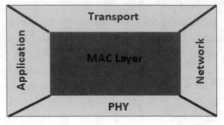

Fig. 1.2 Cross-layer QoS

System efficiency is required to target a mass market in wireless networks, as the QoS requested by a user does not care about the resource utilization and QoS becomes a conflicting concept. However, QoS can be addressed efficiently through a cross-layer design. In this situation, the cross-layer information will be exchanged from the higher layers to the lower ones within the wireless technology protocol stack or vice versa. While in the Open Systems Interconnection (OSI) model [29], nonadjacent layers can only communicate through the intermediate levels, a cross-layer approach can help exchange the trivia between the layers that are not adjacent to each other. Some cross-layer proposals suggest a global coordinator [30, 31], while others recommend a Medium Access Control (MAC)-centric cross-layer air interface, illustrated in Fig. 1.2. The global coordinator gathers the information from the miscellaneous layers and places them in a shared memory. Global coordinator can be in the application or MAC layer.

In a cross-layer approach, the layers have specific requirements. As a case in point, for the PHY, the radio channel should be consistently estimated and the signal strength and the bit error rate (BER) estimations should be available to implement modulation and coding and to select suitable formats at the link layer. For the network layer, mobility should be considered so that the link layer would prioritize the users during handover phases. Moreover, mapping IP QoS features to the link layer radio resource management (RRM) is essential. For transport layer, RRM methods should be modified to treat the Transport Control Protocol (TCP) vs. Universal Datagram Protocol (UDP) [29] and broadcast vs. multicast traffic appropriately. Application layer must consider traffic types and monitor actions performed jointly with the link layer functions to provision with prioritization.

It is notable that this method is protocol-dependent, and if we used an Asynchronous Transfer Mode (ATM) network [29], the cross-layer QoS requirements would be different. For the MAC sublayer, the most important thing is the bandwidth allocation that can be a constraint, e.g. for satellite links in the UL direction shared among the users. Another example for the MAC sublayer is in the context of the ATM networks that need a lookup table including Virtual Path Indicator (VPI) and Virtual Circuit Indicator (VCI) to support the QoS for various ATM service categories. The authors in [32] modified the ATM protocol stack to realize the cross-layer QoS concept. At the network layer, Call Admission Control (CAC) [33] is of high consequence. The CAC algorithms perform a procedure during the call setup if a new connection can be accepted without violating the existing commitments. The CAC methods should be mapped properly into the link layer RRM protocols.

Some other research works concentrated on declaring the QoS requirements through machine understandable languages. For instance, Tournier et al. [34] suggested a component-based QoS architecture using fractional components and considering main QoS concepts. Also, Ahmed et al. [35] proposed an end-to-end QoS model for smart phones using the ontologies. This model addressed basic QoS properties related to the service environment, application services, and user level and leveraged a Web Service Quality Model (WSQM)—an XML-based standardization of expressing the QoS between services and customers—to declare the QoS taxonomy. In essence, their model addressed main elements acting in the service environment (infrastructure, services, and users) and their dynamic nature along with the possibility of incorporating a domain-specific QoS into the modeling. They developed ontologies that could formulate robust QoS descriptions that combined the rich semantics of the QoS ontology with the accuracy of the QoS languages.

The ways to provide with QoS in the IP networks are through Integrated Services (IntServ) [36] and Differentiated Services (DiffServ) [37, 38]. These methods puts QoS mechanisms such as scheduling, routing, and shaping on the routers. IntServ provided with end-to-end QoS guarantees through reserving resources per flow in the nodes along the path by means of Resource Reservation Protocol (RSVP) [39, 40]. The RSVP identifies a session using the destination IP address, port number, and the transport layer protocol for IP version 4 (IPv4) or the destination IP address and flow label for IP version 6 (IPv6). Because reservation needs a policing process, a QoS broker communicates the policy decisions to the routers that may enforce the policies. The RSVP consists of a path message, originating from source, and a reservation message, originating from the destination. The main advantage of this method is provisioning excellent guarantees. Unfortunately, IntServ has severe scalability issues concerning maintaining per-flow operations in the nodes.

In contrast, DiffServ is relatively scalable QoS method; however, it cannot provide any QoS guarantees since it relies on the notion of classifying packets via the differentiated code point (DSCP) field of IP header [41], which defines the three priority classes Best Effort (BE), Assured Forwarding (AF), and Expedited Forwarding (EF). BE provides no guarantees what so ever and its DSCP is 000000. AF subdivides to four classes each with three drop precedence values for the queue. EF whose DSCP is 101110 provides a very small drop probability, latency, and jitter such that it can be considered a virtual leased line (VLL). The EF packets in DiffServ-enabled routers undergo short queues and are quickly serviced. Albeit, DiffServ networks require access control mechanisms in their Access Routers so that only authorized users can inject such high priority packets into the network. Besides QoS-enabled routers, a QoS broker [42] can monitor the network to manage the resources.

1.3   Network Quality of Service Challenges

Because the incessant demand for more data in modern mobile broadband networks grows far beyond the spectrum licensed for commercial wireless communications, there is an urgency to add the spectral resources. While allocating more spectrum to mobile broadband services is very desirable, the current underutilization of the spectrum in the noncommercial domain such as in federal communications encourages a spectrum reuse so as to cater to the mobile broadband demand for bandwidth. This idea of reutilizing spectrum is further motivated by the perpetual bandwidth solicitation from the next-generation mobile broadband networks equipped with interactive traffic-heavy gaming and multimedia applications. Realizing this idea has been proposed by spectrum-governing agencies like FCC through various measures such as spectrum sharing and spectrum auctioning that aim at providing more resources for mobile communications. On the other hand, even though assigning more spectra and reusing under-utilized spectrum for mobile broadband services should alleviate some of the needs of present-day communication networks, smart devices becoming more complex that let them host a wide variety of applications to enhance a wide gamut of daily tasks further diversify the traffic on the network. This traffic diversity brings attention to applications' performance and the users' perception of the network operation, and it can lead to very stringent QoS requirements.

In light of the emerged traffic diversity and QoS constraints, an over-generous addition to broadband bandwidth via spectrum auctions and sharing does not omit the need for developing sophisticated radio resource allocation methods to be an integral part of intelligent next-generation cellular systems. Having a sophisticated resource allocation scheme that can manage temporal data changes and QoS requirements thereof can fine-grain disseminating scarce valuable spectrum efficiently. Such resource allocation schemes should consider a wide range of QoS issues. First, the devised radio resource allocation methods should be efficient so that the resources are not under-utilized and are not wasted. The spectral efficiency means using spectrum such that all the UEs in the RAN receive their required resources based on their current traffic demands. On the other hand, extra resources should not be given to the UEs which could get by much less bandwidth. On the other hand, modern mobile broadband networks are replete with smart devices concurrently running many applications, which range a wide variety of categories to fulfill a large gamut of duties from financial services to educational sessions to entertainment programs. Thus, the applications generate various traffic types with various QoS requirements that should be met so that the applications can perform properly. For example, the delay tolerance of the applications should be considered, and a resource allocation scheme needs to account for the type of the running applications that generate a hybrid traffic with stringent QoS requirements.

Furthermore, the dynamic nature of the cellular communication systems should be taken into the equation. UEs can be highly mobile and move from one area of the RAN to another, and this dynamism should be included in the radio resource allocation. Another dynamism prevalent in modern cellular systems inundated with the smart devices is the temporal variations of the usage percentage of applications running on a smart device. An intelligent resource allocation scheme distributes the resources with temporal usage changes in mind. Moreover, users should be differentiated in present-day cellular systems so that highly prioritized users are treated in a preferential manner. These include public safety responders and national security/emergency preparedness subscribers whose traffic should have the highest possible priority. In addition, users' subscriptions with the MNOs, e.g. prepaid vs. post-paid subscribers, are of the same nature. Also, subscribing to third party services such as Netflix can create a heavy burden on the network and should be included in the novel radio resource allocation approaches in order to ensure fairness in the network and generate revenue for the operators.

While such intelligent spectrum allocation schemes might be derived through mathematically modeling the aforementioned issues, the devised methods should be tailored to the bandwidth assignment units deployed in practical cellular systems, i.e. 4G and 5G. Such cellular technologies rely on a discrete assignment of the spectrum to the UEs, and purely theoretical radio resource allocation techniques may lack the practicality to be applicable to such technologies. Also the devised methods should be computationally efficient as the current volume, and the outgrowth of the mobile Internet traffic as well as gigantic amounts of signaling traffic produced by the mobile broadband is growing and radio resource allocation schemes should assign the resources with a reasonable amount of transmission overhead. Also, how the temporal dynamism of present-day networks from the perspective of the changes induced by the varying number of users in the system and varying application usage percentage in the smart devices would affect the radio resource allocation performance and signaling is important. An excessive signaling can be a limiting factor in realizing a resource allocation technique regardless of the amount of nuances included in its structure.

1.4 Focus of This Book

Because of the fact that the incessant demand for more data in modern-day mobile broadband networks grows far beyond the spectrum licensed for commercial wireless communications, there is an urgency to augment the spectral resources designated for mobile broadband. Even though assigning more spectrum to mobile broadband services is highly desirable, the current federal communications under-utilization of the spectrum puts spectrum reuse as a solution to cater to the mobile broadband's demand for bandwidth. This fascinating notion of spectrum re-utilization is further strengthened by the perpetual bandwidth solicitation from the future mobile broadband networks outfitted with interactive heavy gaming and multimedia applications that can push the data demands exceedingly beyond what network planners can ever envisage. Realization of this idea has been proposed by spectrum-governing agencies like FCC through various measures such as spectrum sharing and spectrum auctioning that aim at provisioning more resources for mobile communications.

On the other hand, while assigning more spectral pieces and reusing under-utilized spectrum for mobile broadband services can alleviate the hunger of present-day communication networks for data to some extent, the evolution of networks' smart devices toward more complexity by letting them host a wide variety of applications to enhance a large gamut of daily tasks further diversifies the traffic types that run on the network. Such traffic diversity is tightly bound to the applications' performance and the users' perception of the network operation, and it can lead to very stringent requirements for the traffic in order to meet the QoS.

In light of the aforesaid traffic diversity and QoS constraints, even over-generously adding to the mobile broadband networks' bandwidth via spectrum auctions and sharing does not eliminate a need for developing sophisticated radio resource allocation entities as integral parts of intelligent cellular communication systems. Having a complex radio resource allocation scheme, handling temporal data demands and QoS requirements of modern communications networks inundated with smart devices, can fine-grain the procedure to disseminate the scarce valuable spectrum efficiently. Such resource allocation schemes should be able to account for a wide range of QoS-related issues. First, the devised radio resource allocation methods should be spectrally efficient in disseminating the available resources so that neither the resources are under-utilized nor they are wasted. The spectral efficiency necessitates that every small portion of the available spectrum is utilized such that all the UEs in the RAN receive their required resources based on their traffic demands. On the other hand, excessive resources should not be assigned to the UEs which require less bandwidth than they are being allocated to.

On the other hand, modern mobile broadband networks are replete with smart devices that can concurrently run many applications to enhance daily tasks. The applications range a wide variety of categories and aim at fulfilling a large gamut of duties from financial services to educational sessions to administrative works to entertainment programs. Therefore, the applications generate various traffic types with miscellaneous QoS requirements that should be met in order for the applications to perform properly. For instance, the tolerance of the applications to the delays in the network should be considered. Therefore, a resource allocation scheme that is to be deployed in modern cellular communication networks should account for the type of the running applications that generate a hybrid traffic with stringent QoS requirements, whose fulfillment elevates the QoE in the network. Furthermore, the dynamic nature of the present-day cellular communication systems should be taken into the equation in the design of the radio resource allocation methods. As an illustration, UEs can be highly mobile and move from one area of the RAN to another, and such a dynamism should be incorporated into the radio resource allocation approaches for the au courant cellular networks.

Another dynamism prevalent in modern cellular communication systems inundated with the smart devices is the temporal changes that occur in the usage percentage of several applications running on a smart device. An intelligent resource allocation scheme that distributes the resources to the smart devices, which run their applications with temporal usage changes, should account for such dynamics in the network in order to efficiently allocate resources in such a time varying system. Moreover, users should be differentiated in present-day cellular communication systems. As highly prioritized users operate in the same cellular communication system as the public users, there should be mechanisms to treat the users in a differential manner. Such high priority users included public safety responders and national security/emergency preparedness subscribers whose traffic should be handled with the highest possible priority. In addition, users can have a variety of subscriptions with the MNOs, e.g. prepaid vs. post-paid subscribers. Besides, subscriptions to third party services can create a heavy burden on the network (e.g. a smart device subscribed to the Netflix). Such subscription-related concerns should be included in the novel radio resource allocation approaches in order to ensure fairness in the network and create revenue for the operators.

While such sophisticated spectrum allocation schemes might be derived through a precise mathematical modeling of all the aforementioned issues, the devised resource allocation methods should be tailored to the bandwidth assignment units deployed in present-day cellular systems, i.e. 4G and 5G. Such cellular technologies rely on a discrete assignment of the spectrum to the UEs, and purely theoretical radio resource allocation techniques may lack the capacity to be applicable to such cellular communication technologies. Additionally, any radio resource allocation scheme should be computationally efficient. Considering the current volume and the outgrowth of the mobile Internet traffic as well as gigantic amounts of signaling traffic generated by the mobile broadband devices, radio resource allocation schemes should be able to assign the resources with a reasonable or possibly minimum amount of transmission overhead. Also, how the temporal dynamism of present-day networks from the perspective of the changes induced by the varying number of users in the system and varying application usage percentage in the smart devices would affect the radio resource allocation performance and signaling is important. An excessive signaling can be a limiting factor in realizing a resource allocation technique regardless of the amount of nuances included in its structure.

Another issue to consider for the radio resource allocation methods, devised for present-day cellular systems and aimed at provisioning QoS in the network, is accommodating bandwidth augmentative novelties such as spectrum sharing. In the light of what was discussed so far, portions of the government-held spectrum [43–47] have been released for mobile broadband purposes. Inasmuch as the majority of the currently released bands are shared with radar systems, novel resource allocation schemes that account for the spectrum sharing with the band incumbents (e.g. radars and satellites) can sustain as a part of the intelligent future-minded cellular communication networks in the long run. Even though the creation of sophisticated yet pragmatic radio resource allocation schemes, which consider the traffic type and dynamics, user subscription, and spectrum sharing, is an effective step toward the development of intelligent cellular communication systems that are capable of meeting the data volume demands and application QoS requirements of the au courant wireless networks, the generated traffic might pass through the core network. Since the core network can contain legacy networks, the voluminous mobile broadband data traffic, treated well in the RAN by an efficient resource management and allocation, can suffer severely in the legacy core network not equipped with sophisticated resource allocation and management schemes. It is noteworthy that while LTE has an all-IP core network, the traffic generated by the smart devices may have to go through legacy networks and there is no way to provision an end-to-end QoS over the entire backhaul networks.

On the other hand, the outrageous perpetual increase in the traffic volume generated in mobile broadband networks causes the backhaul to be a problematic bottleneck in present-day cellular networks. Furthermore, the QoS requirements of the generated traffic will observe grim chances of being fulfilled over the entire backhaul network; this severely adversely impacts the performance of the applications running on the smart devices and degrades the QoE in the cellular network. Such a deterioration of the users' experience, which appears in the form of lengthy delays or excessive losses for the traffic, leads to the subscriber churn. Ultimately, channel conditions for realistic resource allocation are of high consequence. Any realistic resource allocation has to deal with propagation that can put systems in difficult situation in terms of QoS. Hence, a realistic resource allocation should account for channel conditions. Besides, the impact of terrain on propagation is of prominent importance. Hence, equipping resource allocation mechanisms with terrain-aware channel modeling is only meaningful in order to differentiate between users in different terrain conditions in order to assign them resources appropriately.

1.5   Book High Level Contribution and Organization

The contributions of this book are as follows. We develop a radio resource allocation scheme, which is based upon a proportional fairness formulation as well as hybrid real-time and delay-tolerant traffic, for cellular communication systems. Then, we equip the radio resource allocation scheme with an ability to prioritize users, with a capability to prioritize the applications considering their QoS modeling via application utility functions, and with an ability to account for temporal changes of application usages for the UEs present in the system. We prove that the developed radio resource allocation schemes are convex and have a tractable global solution; thereby, the rate allocations achieved through our resource allocation schemes are optimal. Such proof propels the problem of hybrid traffic resource allocation from an (non-deterministic polynomial-time hard) NP-hard problem to one solvable with a polynomial complexity. Besides, we demonstrate that the developed schemes refrain from dropping users by assigning nonzero rates in all times. Moreover, we formulate a centralized architecture for the radio resource allocation scheme and solve it in a single set of message exchanges between UEs and their respective BSs, which indeed directly allocates application rates. Not only do we implement the centralized method on a real-world network, but also we divvy up the application resource allocation into a simpler distributed architecture, which includes network and device optimizations, and provide solution algorithms to give the optimal rates if messages exchanged between the applications and their host UEs and between UEs and their respective BSs. Additionally, we prove the mathematical equivalence of the two architectures. Also, we analyze the transmission overhead of the centralized and distributed approaches as well as the methods' sensitivity to temporal dynamics that occur in the number of UEs or in the application usages in the system. The shadow price convergences for both methods are mathematically analyzed, and a variation of solution algorithm that guarantees with all-time convergence is developed. The radio resource allocation scheme is appended with a channel modeling effect. Terrain-aware propagation modeling is also included to further make the channel-aware resource allocation realistic. Finally, large scale simulations to observe the effect of the channel-aware resource allocation are provided.

The organization of this book is as follows. Chapter 1 presents an introduction to resource allocation in modern communication systems. Chapter 2 discusses the background information needed to understand this book. Specially, it looks at the concept of utility functions and their role in describing QoS of traffic in wireless networks. Chapter 3 presents the resource allocation scheme and provides with solution algorithms thereof. Furthermore, it implements the proposed centralized resource allocation architecture of Chap. 3 on a real-world network and shows that applying the mechanism elevates the QoE in the network. Chapter 4 expands on the resource allocation in Chap. 3 by focusing on traffic analysis of the proposed resource allocation. It further investigates the sensitivity of the proposed architectures to the dynamics incurred in the UE quantity and application usage. Chapter 5 extends the proposed resource allocation framework to account for channel conditions. Chapter 6 presents propagation modeling in order to be leveraged in channel-aware resource allocation formulation of Chap. 5. Chapter 7 presents a large scale simulation that integrates channel-aware resource allocation of Chap. 5 with propagation modeling of Chap. 6. Finally, Chap. 8 concludes the book. This organization is summarized as below.

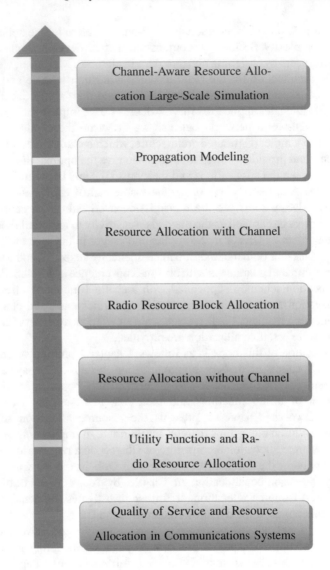

References

1. Mobile data traffic outlook Ericsson. https://www.ericsson.com/en/mobility-report/reports/june-2020/mobile-data-traffic-outlook. Accessed 10 Nov 2020
2. M. Ghorbanzadeh, A. Abdelhadi, C. Clacy, *Cellular Communications Systems in Congested Environments Resource Allocation and End-to-End Quality of Service Solutions with MATLAB* (Springer, Berlin, 2017)

3. M. Ghorbanzadeh, A. Abdelhadi, C. Clancy, A utility proportional fairness bandwidth allocation in radar-coexistent cellular networks, in *Military Communications Conference (MILCOM)*, 2014

4. Y. Chen, M. Ghorbanzadeh, K. Ma, C. Clancy, R. McGwier, A hidden Markov model detection of malicious Android applications at runtime, in *2014 23rd Wireless and Optical Communication Conference (WOCC)*, 2014

5. G. Intelligence, Smartphone users spending more 'face time' on apps than voice calls or web browsing. Technical Report, 2011

6. N. S. Networks, Understanding smartphone behavior in the network. White Paper, 2011

7. M. Ghorbanzadeh, Resource allocation and end-to-end quality of service for cellular communications systems in congested and contested environments, in Ph.D. Thesis, Virginia Tech, 2015

8. A. Ghosh, R. Ratasuk, Essentials of LTE and LTE-A. The Cambridge Wireless Essentials Series, 2011

9. M. Ghorbanzadeh, Y. Chen, K. Ma, C. Clancy, R. McGwier, A neural network approach to category validation of Android applications, in *IEEE Conference on Computing, Networking, and Communications (ICNC)*, 2013

10. G. Piro, L. Grieco, G. Boggia, P. Camarda, A two-level scheduling algorithm for QoS support in the downlink of LTE cellular networks, in *Wireless Conference (EW)*, 2010

11. M. Marchese, M. Mongelli, Reference chaser bandwidth controller for wireless QoS mapping under delay constraints. EURASIP J. Wirel. Commun. Netw. (2010)

12. D. Soldani, H.X. Jun, B. Luck, Strategies for mobile broadband growth: traffic segmentation for better customer experience. *IEEE Vehicular Technology Conference (VTC)*, 2011

13. A. Larmo, M. Lindstrom, M. Meyer, G. Pelletier, J. Torsner, H. Wiemann, The LTE link-layer design. IEEE Commun. Mag. **47**(4). 52—59 (2009)

14. C. Ciochina, H. Sari, A review of OFDMA and single-carrier FDMA, in *Wireless Conference (EW)*, 2010

15. S. Ali, M. Zeeshan, A delay-scheduler coupled game theoretic resource allocation scheme for LTE networks, in *Frontiers of Information Technology (FIT)*, 2011

16. D. Fudenberg, J. Tirole, *Nash Equilibrium: Multiple Nash Equilibria, Focal Points, and Pareto Optimality* (MIT Press, Cambridge, 1991)

17. P. Ranjan, K. Sokol, H. Pan, Settling for less—a QoS compromise mechanism for opportunistic mobile networks, in *SIGMETRICS Performance Evaluation*, 2011

18. R. Johari, J. Tsitsiklis, Parameterized supply function bidding: equilibrium and efficiency. Oper. Res. **59**(5), 1079-1089 (2011)

19. H. Yin, S. Alamouti, OFDMA: a Broadband Wireless Access Technology, in *IEEE Sarnoff Symposium*, 2006

20. L. Chung, Energy efficiency of QoS routing in multi-hop wireless networks, in *IEEE International Conference on Electro/Information Technology (EIT)*, 2010

21. H. Ekstrom, QoS control in the 3GPP evolved packet system. IEEE Commun. Mag. **47**(2), 76–83 (2009)

22. G. T. . V9.0.0, Further advancements for E-UTRA physical layer aspects, in *Measuring of Heterogeneous Wireless and Wired Networks*, 2012

23. M. Alasti, B. Neekzad, H. Jie, R. Vannithamby, Quality of service in WiMAX and LTE networks [Topics in Wireless Communications]. IEEE Commun. Mag. **48**(5), 104–111 (2010)

24. J. Andrews, A. Ghosh, and R. Muhamed, "Fundamentals of WiMAX: Understanding broadband wireless networking," Prentice Hall Communications Engineering and Emerging Technologies Series, 2007.

25. B. Mota, Quality of service in wireless backhaul applications with VortiQa software for service provider equipment, in *Freescale*, 2010

26. IXIACOM, Quality of Service (QoS) and Policy Management in Mobile Data Networks. White Paper, 2010

27. F. Li, Quality of service, traffic conditioning, and resource management in universal mobile telecommunication system (UMTS). Doctoral Dissertation, Norwegian University of Science and Technology, 2003
28. G. Gorbil, I. Korpeoglu, Supporting QoS traffic at the network layer in multi-hop wireless mobile networks, in *Wireless Communications and Mobile Computing Conference (IWCMC)*, 2011
29. W. Stallings, Data and computer communications, in *William Stallings Books on Computer and Data Communications*, 2013
30. C. Dovrolis, D. Stiliadis, P. Ramanathan, Proportional differentiated services: delay differentiation and packet scheduling, in *IEEE/ACM Transactions on Networking*, 2002
31. A. Sali, A. Widiawan, S. Thilakawardana, R. Tafazolli, B. Evans, Cross-layer design approach for multicast scheduling over satellite networks, in *2nd International Symposium on Wireless Communication Systems, 2005*, 2005
32. E. Lutz, D. Cygan, M. Dippold, F. Dolainsky, W. Papke, The land mobile satellite communication channel-recording, statistics, and channel model. IEEE Trans. Veh. Technol. **40**(2), 375–386 (1991)
33. H. Perros, K. Elsayed, Call admission control schemes: a review. IEEE Commun. Mag. **34**(11), 82–91 (1996)
34. J. Tournier, J. Babau, V. Olive, Qinna, a component-based QoS architecture, in *Proceedings of the 8th International Conference on Component-Based Software Engineering*, 2005
35. N. Ahmed, H. Yan, Access control for MPEG video applications using neural network and simulated annealing, in *Mathematical Problems in Engineering*, 2004
36. R. Braden, Integrated services in the internet architecture: an overview. IETF RFC 1633, 1994
37. S. Blake, An architecture for differentiated services. IETF RFC 2475, 1998
38. M. Ghorbanzadeh, Y. Chen, C. Clancy, Fine-grained end-to-end network model via vector quantization and hidden Markov processes, in *IEEE Conference on Communications (ICC)*, 2013
39. R. Braden, Resource ReServation Protocol (RSVP)—version 1 functional specification. IETF RFC 2205, 1997
40. M. Ghorbanzadeh, A. Abdelhadi, C. Clancy, A utility proportional fairness radio resource block allocation in cellular networks, in *IEEE International Conference on Computing, Networking and Communications (ICNC)*, 2015
41. K. Nichols, A two-bit differentiated services architecture for the internet. IETF RFC 2638, 1999
42. K. Nahrstedt, The QoS broker, *IEEE Multimedia*, 1995
43. M. Ghorbanzadeh, E. Visotsky, P. Moorut, W. Yang, C. Clancy, Radar inband and out-of-band interference into LTE macro and small cell uplinks in the 3.5 GHz band, in *2015 IEEE Wireless Communications and Networking Conference (WCNC)*, 2015
44. M. Ghorbanzadeh, E. Visotsky, P. Moorut, W. Yang, C. Clancy, Radar in-band interference effects on macrocell LTE uplink deployments in the U.S. 3.5 GHz band, in *2015 International Conference on Computing, Networking and Communications (ICNC)*, 2015
45. M. Ghorbanzadeh, E. Visotsky, P. Moorut, W. Yang, C. Clancy, Radar interference into LTE base stations in the 3.5 GHz band. Phys. Commun. **20**, 33–47 (2016)
46. H. Shajaiah, M. Ghorbanzadeh, A. Abdelhadi, C. Clancy, Application-aware resource allocation based on channel information for cellular networks, in *2019 IEEE Wireless Communications and Networking Conference (WCNC)* (2019), pp. 1–6
47. M. Ghorbanzadeh, A. Abdelhadi, C. Clancy, Application-aware resource allocation of hybrid traffic in cellular networks. IEEE Trans. Cogn. Commun. Netw. **3**(2), 226–241 (2017)

Chapter 2
Utility Functions and Radio Resource Allocation

2.1 Introduction

RF resource allocation in wireless communications has been studied extensively in research works with methods from linear algebra, queuing theory, machine learning, and so forth. Modeling QoS-based RF resource allocation as optimization of utility functions that define traffic QoS requirements has been a major focus among research efforts, and an understanding of utility functions modeling traffic QoS and optimization theory methods is essential to grasp RF resource allocation methods of this book. To this end, this chapter presents the background information on the aforesaid subjects to the extent necessary to facilitate absorption of material at hand. Furthermore, a succinct literature survey to familiarize the reader with the gist of RF resource allocation methods is presented in Sect. 2.2 of this chapter. In this chapter, we explain the following matters.

- The background information on application utility functions as a QoS modeling measure is provided.
- Popular RF resource allocation formulations of cellular communication systems are briefly reviewed.
- A deeper dive into utility proportional fairness formulation of RF resource allocation is explained.
- Mechanisms to differentiate traffic based on the QoS requirements, application usage changes in time, and user priorities are elaborated.

The rest of this chapter proceeds as follows. Section 2.2 illustrates a literature survey on resource allocation in modern cellular communication systems, Sect. 2.3 introduces application utility functions, Sect. 2.4 presents common proportional fairness resource allocation formulation, and Sect. 2.5 summarizes this chapter.

© The Author(s), under exclusive license to Springer Nature Switzerland AG 2022
M. Ghorbanzadeh, A. Abdelhadi, *Practical Channel-Aware Resource Allocation*,
https://doi.org/10.1007/978-3-030-73632-3_2

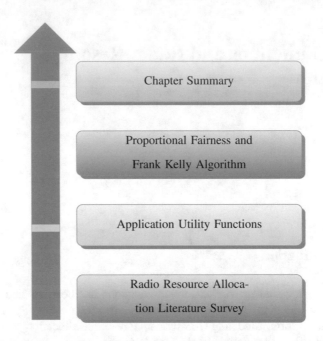

2.2   Radio Resource Allocation Literature Survey

The radio resource allocation optimization research has received significant interest after the network utility maximization work in [1] that assigned UE rates by a utility proportional fairness maximization that is solved by means of Lagrange multipliers [2]. Soon after this study, an iterative solution algorithm using the optimization duality was proposed [3–5]. The traffic characteristics of these research works have an elastic nature, suitable for wired communication systems, and are modelled by concave utility functions. Nonetheless, modern high-speed wireless networks have increased the usage of real-time applications with non-concave application utility functions [6]. For instance, VoIP application utility functions can be represented as a step function whose utility becomes 0 and 100% before and after a minimum threshold rate is met. Another example is a video streaming application with sigmoidal application utility functions, convex/concave for rates below/above its inflection point. As such, the approaches presented in [1, 3] have the following drawbacks that (i) reaching an optimal solution for concave utility functions, these methods are not applicable to the ever-more common inelastic traffic of modern networks, (ii) they do not prioritize real-time applications with stringent QoS requirements, (iii) they do not provide any attention to the application usage dynamics in time, and (iv) they cannot prioritize subscribers who can be pivotal from a business and/or security standpoint.

Next, the authors in [7, 8] introduced distributed resource allocation algorithms that use concave and sigmoidal application utility functions, but in spite of approximating an optimal solution, the methods could drop UEs in order to maximize the system utility; thereby, it could not guarantee a minimal QoS. The authors in [9–11] proposed a utility proportional fairness resource allocation for a wireless communication system as a convex optimization with logarithmic and sigmoidal utility functions modeling delay-tolerant and real-time applications, respectively. While their approach prioritized the real-time applications over the delay-tolerant applications, application status, UE prioritization treatments, and now-common hybrid traffic were not considered.

The authors in [12, 13] and [14–16] proposed a similar multi-carrier optimal resource allocation with subscriber prioritization with no heeding to the application usage changes in time or UE quantities. In [17], the authors used a non-convex optimization to maximize the resource allocation system utility containing logarithmic and sigmoidal application utility functions. A distributed procedure was used to obtain the assigned rates under a zero duality gap, yet the algorithm could not converge for a positive duality gap that led to a heuristic approach to ensure network stability. In [18], the author considered a weighted aggregation of logarithmic and sigmoidal application utility functions approximated to the nearest concave utility function via a minimum mean-squared error (MMSE) measure inside UEs. The approximate application utility function solves the rate allocation optimization through a variation of the conventional distributed resource allocation method in [1] such that rate assignments estimate optimal ones. This work was expanded by Shajaiah et. al. [19, 20] to allow for a multi-carrier network.

In a similar work [21–23], the authors used a utility proportional fairness optimization that assigned optimal UE rates in a cellular spectrum sharing ecosystem coexistent with radar systems [24–27] presented a subcarrier level allocation in orthogonal frequency division multiplexed (OFDM) systems and leveraged network delay models [28–30] for subcarrier assignments. The authors of [31, 32] developed a utility max–min fairness resource allocation for the hybrid traffic sharing a single path of a communication network. Similarly, [33–35] presented a utility proportional fairness optimization for systems with a high signal-to-interference-plus-noise ratio (SINR) via a utility max–min formulations, contrasted against the proportional fairness methods [36–39], and gave a closed-form solution that eliminated network oscillations. However, the methods did not pay any attention to traffic type or user priorities in assigning the rates [40–42] created a utility proportional fairness resource block allocation as an integer optimization problem by obtaining the continuous optimal rates first followed by a boundary-mapping technique to extract a set of resource blocks equivalent to the optimal continuous rates.

Bjornson has extensively written on radio resource allocation [43–51], and many theoretical contributions are presented in [8, 52–61, 61–63]. Finally, [64] proposed a context-aware source allocation in cellular networks that did not consider the temporal changes of the application usage percentages, the quantity of UEs in the system, or the subscriber prioritization.

2.3   Application Utility Functions

Application utility functions are used to model the characteristic features of the system such as QoS as shown in [28, 65–69]; these works leveraged utility functions for modeling modulation schemes of power allocation. They can additionally be leveraged in sensors for optimal Machine-to-Machine Communications (M2MC) [70–75]. For modern cellular networks, applications on the smart devices have QoS requirements whose fulfillment can be expressed by application utility functions showing feasible rates for QoS fulfillment percentage. Traditional applications such as File Transfer Protocol (FTP) and Simple Mail Transfer Protocol (SMTP) are delay-tolerant and have elastic traffic adaptable in facing congestion and network delay [21, 40, 76–80]. Intuitively, the larger the allocated rate of the application, the higher the QoS satisfaction for the application utility function.

The application utility function modeling of the QoS satisfaction of the elastic traffic from delay-tolerant applications looks like Fig. 2.1 where a slight increase of the value of the application utility function, as the allocated rate is increased, is observed. Another observation is that the logarithmic application utility function in Fig. 2.1 is convex. This property proves integral to easy solving of resource

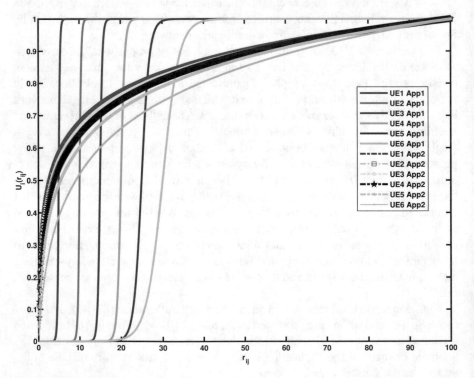

Fig. 2.1 6 UEs each running a delay-tolerant and a real-time application with identically colored logarithmic and sigmoidal utility functions U_{ij} vs. rates r_{ij}

allocation as we will observe later in this chapter. On the other hand, applications like telephony and video conferencing generate an inelastic traffic that requires minimum QoS requirement rate. Such applications, which perform poorly if a minimum data rate needed to meet a certain delay bound is not met, are referred to as real-time applications. They ask for a minimum throughput before achieving an acceptable performance and their application utility function is shown in Fig. 2.1, which depicts that, after meeting a minimum rate, the performance satisfaction remains almost constant whereas a throughput dropping below a minimum rate needed to meet the required delay bounds slashes down the performance to zero abruptly. While audio and video (AV) applications have hard QoS requirements, they are rather tolerant of occasional delay/packet loss. Hence, the QoS performance degrades severely as the application rate grows less than the intrinsic data generation rate for that traffic and leads to a soft real-time application. For these application utility functions, the performance satisfaction percentage remains mainly constant after meeting the minimum rate, whereas it plummets close to zero when the application allocated rate gets less than the value to meet the delay bounds. In comparison to hard real-time applications, the performance decline of the delay-adaptive real-time application is not as drastic as the hard real-time application shown in Fig. 2.1, which shows that the utility function shape is convex and concave, respectively, before and after the minimum rate for the application utility function. Hereafter, unless it is explicitly stated, this book refers to an application performance satisfaction as a function of rate as application utility function, denoted as $U(r)$ for the allocated rate r. Application utility functions' main properties are that $U(0) = 0$ and $U(r)$ is an increasing function r, $U(r)$ is twice differentiable in r, and $U(r)$ is upper-bounded [6, 81–85].

The first statement implies the connectivity of the application utility functions and is expected since they represent application performance satisfaction. The second and third statements reveal that the larger the rates, the higher the QoS satisfaction and indicate that the application utility functions are continuous and bounded. A hybrid traffic contains both inelastic and elastic traffic streams from real-time and delay-tolerant applications with QoSs modeled by sigmoidal and logarithmic utility functions given in Eqs. (2.2) and (2.1), respectively [7, 86].

$$U(r) = c\left(\frac{1}{1 + e^{-a(r-b)}} - d\right) \tag{2.1}$$

Here, $c = \frac{1+e^{ab}}{e^{ab}}$ and $d = \frac{1}{1+e^{ab}}$. It can be verified that $\lim_{r \to \infty} U(r) = 1$ and $U(0) = 0$, where the first equality indicates that an infinite resource assignment leads to 100% satisfaction and the second equality is a repetition of previously mentioned application utility functions. Moreover, it can be easily derived that the application utility function inflection point in Eq. (2.1) is at $r = r^{\text{inf}} = b$, where the superscript "inf" abbreviates "infliction." This inflection can be done by differentiating $U(r)$ with respect to r twice and setting the second derivative equal to zero as $\frac{\partial^2 U}{\partial r^2} = 0 \rightarrow r = b$. Here, r^{max} is the maximum rate for which the

Table 2.1 Application utility parameters

Application utility parameters	
UE1 App1, App2	Sigmoidal $a = 5$, $b = 5$, Logarithmic $k = 15$, $r^{\max} = 100$
UE2 App1, App2	Sigmoidal $a = 4$, $b = 10$, Logarithmic $k = 12$, $r^{\max} = 100$
UE3 App1, App2	Sigmoidal $a = 3$, $b = 15$, Logarithmic $k = 9$, $r^{\max} = 100$
UE4 App1, App2	Sigmoidal $a = 2$, $b = 20$, Logarithmic $k = 6$, $r^{\max} = 100$
UE5 App1, App2	Sigmoidal $a = 1$, $b = 25$, Logarithmic $k = 3$, $r^{\max} = 100$
UE6 App1, App2	Sigmoidal $a = 0.5$, $b = 30$, Logarithmic $k = 1$, $r^{\max} = 100$

application QoS is fully satisfied ($U(r^{\max}) = 100\%$) and k is the application utility function increase as the rate r also goes up. The normalized logarithmic function inflection point occurs at $r = r^{\inf} = 0$.

$$U(r) = \frac{\log(1 + kr)}{\log(1 + kr^{\max})} \qquad (2.2)$$

The facts that sigmoidal application utility functions gain a small QoS satisfaction after the allocated rates exceed the inflection points and logarithmic utility functions enjoy some QoS fulfillment for even a small rate increase make sigmoidal and logarithmic application utility functions representative models for real-time and delay-tolerant traffic QoS satisfaction respectively. Mathematical analyses related to this matter appear in [7, 33, 87, 88] in more detail. Consider 6 UEs each concurrently running a delay-tolerant and a real-time application modeled by logarithmic and sigmoidal application utility functions where the sigmoidal utility parameters $a = 5$ and $b = 10$ approximate a step function at rate $r = 5$, parameters $a = 3$ and $b = 15$ are an approximation of a real-time application with an inflection point at rate $r = 15$, parameters $a = 1$ and $b = 25$ are estimations of real-time applications with the inflection point at $r = 25$, and logarithmic utilities with $r^{\max} = 100$ and distinct k_i parameters represent delay-tolerant applications. The plots of application utility functions are shown in Fig. 2.1 that affirms that utility functions are increasing and zero valued at zero rates. The parameters are summarized in Table 2.1.

2.3.1 Application Utility Function MATLAB Code

The following code creates and plots application utility functions, given in Fig. 2.1. The code lines 1–12 initialize the application utility function parameters according to Table 2.1 and require MATLAB Symbolic toolbox. If the reader does not have access to the toolbox, you can declare x in the code as a vector such as $x = 0.1 :$ 1000; and eliminate "syms x."

```
1   %Create application utility functions
2   syms x
3   k = 15;
4   a = 1;
5   b = 25;
6   c = (1 + exp(a .* b)) ./ (exp(a .* b));
7   d = 1 ./ (1 + exp(a .*b));
8   for i = 1: length(a)
9       % Sigmoid utility function
10      y(i) = c(i).*(1./(1+exp(-a(i).*(x-b(i))))-d(i));
11      % Logarithmic utility function
12      y2(i) = log(k(i).*x+1)./(log(k(i).*100+1));
13      m(i) = exp(-a(i).*(x-b(i)));
14      dy_log(i) = k(i)./((1+k(i).*x).*log(1+k(i).*x));
15  end
16  %Plot application utility functions
17  yy(1) = 0;
18  yy2(1) = 0;
19  for j = 2:1: 1000
20      x0(j) = 0.1 * j ;
21      yy(j) = subs(y(i),x0(j));
22      yy2(j) = subs(y2(i),x0(j));
23  end
24  plot(x0,yy,x0,yy2);
25  legend('Sigmoidal', 'Logarithmic');
```

2.4 Proportional Fairness and Frank Kelly Algorithm

A wide variety of formulations have been presented in the mathematics literature, relevant to the use of application utility functions for resource allocation, which also subsume a variety of solutions to the aforesaid resource allocation formulations among which max–min [89–96] and proportional fairness formulations [97] have received attention since they produce optimal and/or efficient solutions [98–100]. Moreover, the authors of [18, 101, 102] define the concepts of Pareto inefficient, Pareto optimal, and infeasible solutions. The third term indicates a rate allocation is not feasible in view of the available resources and the network demand. However, [18] defined Pareto inefficient solution as a rate assignment that does not allocate resources; it further defines Pareto optimality as allocating all the available resources. This and next chapters of this book will leverage formulations that lead to Pareto optimal solutions.

A feasible resource allocation is proportionally fair if it maximizes the overall system utility while providing with a minimal service to individual system entities needing resources [67]. This proportionally fair resource allocation is performed by allocating each entity a rate inversely proportional to its resource needs [103, 104]. For the application utility function $U_i(r_i)$, where r_i is the rate assigned to the ith UE, proportional fairness formulation can be written as Eq. (2.3). The application utility functions' properties in Sect. 2.3 included $U_i(r_i = 0) = 0$ zeroing the system utility ($\prod_{i=1}^{N} U_i(r_i)$); hence, no UE is assigned a zero rate with this formulation. Various methods for solving proportional fairness optimizations have been proposed in the literature such as Weighted Fair Queuing (WFQ) [105–107] and Frank Kelly algorithm [1]. The latter is an iterative algorithm that allows UEs to bid for resources until the algorithm achieves an optimal rate allocation and the shadow price, number of consumed resources per data bit [1]. On the other hand, proportional fairness resource allocation can be obtained by setting the inverse shadow price as the weights used for the WFQ. We will leverage a proportional fairness resource allocation formulation in this and next chapter upon which we built more and more to include UE prioritization and application temporal usage.

$$r_i = argmax_{r_i} \prod_{i=1}^{N} U_i(r_i) \tag{2.3}$$

Frank Kelly algorithm, a seminal work realizing proportional fairness resource allocation, was introduced in [1], which proved the Pareto optimal solution of a proportional fairness resource allocation formulation. Frank Kelly algorithm starts by the UEs to send their bids w_i to a resource allocation entity that calculates a shadow price, defined as the addition of the bids averaged on the total resources R available at the resource allocation entity, i.e., $p = \frac{\sum_i^N w_i}{R}$. The bid to shadow price ratio, $r_i = \frac{w_i}{p}$, derives the assigned rates. Next, UEs evaluate if the assigned rate is optimal by solving $r_{i,textopt} = \arg\max_{r_i}\left(U_i(r_i) - pr_i\right)$ and if $r_i \neq r_{i,\text{opt}}$; then, they send their new bids $w_i = r_{i,\text{opt}} p$ to the resource allocation entity. The procedure iterates until convergence by equating the application utility function derivative to the shadow price $\frac{\partial U_i}{\partial r_i}|_{r_i = r_{i,\text{opt}}} = p$ [1, 18]. This entire procedure is summarized in Algorithm 1. In the current and future chapters, a method based on the Frank Kelly algorithm is employed to solve the proportional fairness resource allocation formulation. Next, Sect. 2.2 presents a short literature survey on radio resource allocation.

2.5   Chapter Summary

This chapter introduced the concept of application utility functions for modeling QoS requirement of delay-tolerant and real-time applications via logarithmic and sigmoidal utility functions, respectively. This chapter explained max–min and proportional fairness optimizations as common approaches to resource allocation and developed a QoS-minded utility proportional fairness resource allocation, for delay-tolerant and real-time applications mathematically modeled as logarithmic and sigmoidal application utility functions correspondingly.

Algorithm 1 Frank Kelly algorithm

Send initial bid $w_i(n = 1)$ to the resource allocation entity.
loop

　Compute shadow price $p(n) = \frac{\sum_i^N w_n}{R}$.
　Receive shadow price $p(n)$ from the resource allocation entity.
　Calculate allocated rate $r_i = \frac{w_i(n)}{p(n)}$.
　Solve $r_{i,\text{opt}} = \arg\max_{r_i}\Big(U_i(r_i) - p(n)r_i\Big)$.
　if $r_i \neq r_{i,\text{opt}}$ **then**
　　Calculate $w_i = r_{i,\text{opt}} p$.
　　Send new bid $w_i(n)$ to resource allocation entity.
　end if
end loop

References

1. F. Kelly, A. Maulloo, D. Tan, Rate control in communication networks: shadow prices, proportional fairness and stability. J. Oper. Res. Soc. **49**(3), 237–252 (1998)
2. S. Boyd, L. Vandenberghe, *Introduction to Convex Optimization with Engineering Applications* (Cambridge University Press, Cambridge, 2004)
3. S. Low, D. Lapsley, Optimization flow control, I: basic algorithm and convergence. IEEE/ACM Trans. Netw. **7**(6), 861–874 (1999)
4. P. de Kerret, D. Gesbert, The multiplexing gain of the network MIMO channel with distributed CSI, in *2011 IEEE International Symposium on Information Theory Proceedings* (IEEE, Piscataway, 2011)
5. M. Ghorbanzadeh, A. Abdelhadi, C. Clancy, A utility proportional fairness resource allocation in spectrally radar-coexistent cellular networks, in *Military Communications Conference (MILCOM)*, 2014
6. S. Shenker, Fundamental design issues for the future Internet. IEEE J. Sel. Areas Commun. **13**(7), 1176–1188 (1995)
7. J. Lee, R. Mazumdar, N. Shroff, Downlink power allocation for multi-class wireless systems. IEEE/ACM Trans. Netw. **13**(4), 854–867 (2005)
8. H. aShajaiah, A. Abdelhadi, C. Clancy, Spectrum sharing approach between radar and communication systems and its impact on radar's detectable target parameters, in *2015 IEEE 81st Vehicular Technology Conference (VTC Spring)* (2015), pp. 1–6
9. A. Abdelhadi, C. Clancy, A utility proportional fairness approach for resource allocation in 4G-LTE, in *IEEE International Conference on Computing, Networking, and Communications (ICNC), CNC Workshop*, 2014
10. A. Abdelhadi, C. Clancy, A robust optimal rate allocation algorithm and pricing policy for hybrid traffic in 4G-LTE, in *IEEE International Symposium on Personal, Indoor, and Mobile Radio Communications (PIMRC)*, 2013
11. A. Abdelhadi, C. Clancy, J. Mitola, A resource allocation algorithm for users with multiple applications in 4G-LTE," in *ACM Workshop on Cognitive Radio Architectures for Broadband (MobiCom Workshop CRAB)*, 2013
12. M. Han, T. Yu, J. Kim, K. Kwak, S. Lee, S. Han, D. Hong, OFDM channel estimation with jammed pilot detector under narrow-band jamming. IEEE Trans. Veh. Technol. **57**(3), 1934–1939 (2008)
13. M. Ghorbanzadeh, A. Abdelhadi, T. Clancy, A utility proportional fairness approach for resource block allocation in cellular networks, in *IEEE International Conference on Computing, Networking and Communications (ICNC)*, 2015

14. H. Shajaiah, A. Abdelhadi, C. Clancy, Multi-application resource allocation with users discrimination in cellular networks, in *IEEE International Symposium on Personal, Indoor and Mobile Radio Communications*, 2014
15. Y. Yu, A jamming scheme based on pilot assisted channel estimation of OFDM. J. Electron. Inf. Warfare Technol. (2008)
16. A. Khawar, A. Abdel-Hadi, T.C. Clancy, MIMO radar waveform design for coexistence with cellular systems, in *IEEE International Symposium on Dynamic Spectrum Access Networks: SSPARC Workshop (IEEE DySPAN 2014—SSPARC Workshop)*, 2014
17. G. Tychogiorgos, A. Gkelias, K. Leung, A new distributed optimization framework for hybrid ad-hoc networks, in *GLOBECOM Workshops*, 2011
18. R. Kurrle, Resource allocation for smart phones in 4G-LTE advanced carrier aggregation. Master Thesis, Virginia Tech, 2012
19. H. Shajaiah, A. Abdelhadi, C. Clancy, Spectrum sharing between public safety and commercial users in 4G-LTE, in *IEEE International Conference on Computing, Networking and Communications (ICNC)*, 2014
20. H. Shajaiah, A. Abdelhadi, C. Clancy, Utility proportional fairness resource allocation with carrier aggregation in 4G-LTE, in *IEEE Military Communications Conference (MILCOM)*, 2013
21. M. Ghorbanzadeh, A. Abdelhadi, C. Clancy, A utility proportional fairness bandwidth allocation in radar-coexistent cellular networks, in *Military Communications Conference (MILCOM)*, 2014
22. A. Khawar, A. Abdel-Hadi, T.C. Clancy, Spectrum sharing between S-band radar and LTE cellular system: a spatial approach, in *2014 IEEE International Symposium on Dynamic Spectrum Access Networks: SSPARC Workshop (IEEE DySPAN 2014—SSPARC Workshop)*, 2014
23. M. Ghorbanzadeh, A. Abdelhadi, A. Amanna, J. Dwyer, T. Clancy, Implementing an optimal rate allocation tuned to the user quality of experience, in *2015 International Conference on Computing, Networking and Communications (ICNC)* (2015), pp. 292–297
24. T. Jiang, L. Song, Y. Zhang, Orthogonal frequency division multiple access fundamentals and applications. Auerbach Publications, 2010
25. M. Dohler, R. Heath, A. Lozano, C. Papadias, R. Valenzuela, Is the PHY layer dead? IEEE Commun. Mag. **49**(4), 159–165 (2011)
26. Y. Eldar, N. Merhav, A competitive minimax approach to robust estimation of random parameters. IEEE Trans. Signal Process. **52**(7), 1931–1946 (2004)
27. K. Eriksson, S. Shi, N. Vucic, M. Schubert, E. Larsson, Globally optimal resource allocation for achieving maximum weighted sum rate, in *IEEE Global Communications Conference*, 2010
28. M. Ghorbanzadeh, Y. Chen, C. Clancy, Fine-grained end-to-end network model via vector quantization and hidden Markov processes, in *IEEE Conference on Communications (ICC)*, 2013
29. P. de Kerret, D. Gesbert, Towards optimal CSI allocation in multicell MIMO channels, in *Proceedings of IEEE International Conference on Communications*, 2012
30. H. Shajaiah, A. Abdelhadi, T.C. Clancy, A price selective centralized algorithm for resource allocation with carrier aggregation in LTE cellular networks, in *2015 IEEE Wireless Communications and Networking Conference (WCNC)* (IEEE, Piscataway, 2015), pp. 813–818
31. T. Harks, Utility proportional fair bandwidth allocation: an optimization oriented approach, in *QoS-IP*, 2005
32. A. Abdelhadi, C. Clancy, Context-aware resource allocation in cellular networks (2014). arXiv preprint arXiv:1406.1910
33. G. Tychogiorgos, A. Gkelias, K. Leung, Utility proportional fairness in wireless networks, in *IEEE International Symposium on Personal, Indoor, and Mobile Radio Communications (PIMRC)*, 2012
34. R. Etkin, A. Parekh, D. Tse, Spectrum sharing for unlicensed bands. IEEE J. Sel. Areas Commun. **25**(3), 517–528 (2007)

35. R. Etkin, D. Tse, and H. Wang, "Gaussian interference channel capacity to within one bit," in *IEEE Transactions on Information Theory*, 2008.
36. T. Nandagopal, T. Kim, X. Gao, V. Bharghavan, Achieving MAC layer fairness in wireless packet networks, in *Proceedings of the 6th annual International Conference on Mobile Computing and Networking (MobiCom)*, 2000
37. P. Dighe, R. Mallik, S. Jamuar, Analysis of transmit-receive diversity in Rayleigh fading. IEEE Trans. Commun. **51**(4), 694–703 (2003)
38. P. Dighe, R. Mallik, S. Jamuar, On downlink beamforming with greedy user selection: performance analysis and a simple new algorithm. IEEE Trans. Signal Process. **53**(10), 3857–3868 (2005)
39. M. Ghorbanzadeh, A. Abdelhadi, T. Clancy, A utility proportional fairness bandwidth allocation in radar-coexistent cellular networks, in *Military Communications Conference (MILCOM)*, 2014
40. M. Ghorbanzadeh, A. Abdelhadi, C. Clancy, A utility proportional fairness radio resource block allocation in cellular networks, in *IEEE International Conference on Computing, Networking and Communications (ICNC)*, 2015
41. T. Erpek, A. Abdelhadi, C. Clancy, An optimal application-aware resource block scheduling in LTE, in *IEEE International Conference on Computing, Networking and Communications (ICNC) (Worshop CCS)*, 2015
42. A. Abdelhadi, C. Clancy, An optimal resource allocation with joint carrier aggregation in 4G-LTE, in *2015 International Conference on Computing, Networking and Communications (ICNC)* (2015), pp. 138–142
43. E. Bjornson, Multiantenna cellular communications: Channel estimation, feedback, and resource allocation, in Ph.D. thesis, KTH Royal Institute of Technology, 2011
44. E. Bjornson, M. Bengtsson, B. Ottersten, Receive combining vs. multistream multiplexing in multiuser MIMO systems, in *in Proceedings of IEEE Swedish-Communication Technologies Workshop*, 2001
45. E. Bjornson, M. Bengtsson, B. Ottersten, Pareto characterization of the multicell MIMO performance region with simple receivers, in *IEEE Transactions on Signal Processing*, 2012
46. E. Bjornson, M. Bengtsson, B. Ottersten, Computational framework for optimal robust beamforming in coordinated multicell systems, in *Proceedings of IEEE Computational Advance in Multi-Sensor Adaptive Processing*, 2011
47. E. Bjornson, M. Bengtsson, B. Ottersten, Optimality properties, distributed strategies, and measurement-based evaluation of coordinated multicell OFDMA transmission, in *IEEE Transactions on Signal Processing*, 2011
48. E. Bjornson, B. Ottersten, A framework for training-based estimation in arbitrarily correlated Rician MIMO channels with Rician disturbance. IEEE Trans. Signal Process. **58**(3), 1807–1820 (2009)
49. E. Bjornson, R. Zakhour, D. Gesbert, B. Ottersten, Cooperative multicell precoding: rate region characterization and distributed strategies with instantaneous and statistical CSI. IEEE Trans. Signal Process. **58**(8), 4298–4310 (2010)
50. E. Bjornson, R. Zakhour, D. Gesbert, B. Ottersten, Optimal coordinated beamforming in the multicell downlink with transceiver impairments, in *IEEE Global Communication Conference*, 2012
51. E. Bjornson, R. Zakhour, D. Gesbert, B. Ottersten, Characterization of convex and concave resource allocation problems in interference coupled wireless systems. IEEE Trans. Signal Process. 59(5), 2382–2394 (2011)
52. H. Boche, M. Schubert, A general duality theory for uplink and downlink beamforming, in *IEEE Vehicular Technology Conference*, 2002
53. H. Boche, M. Schubert, A calculus for log-convex interference functions. IEEE Trans. Inf. Theory 54(12), 5469–5490 (2008)
54. F. Bock, B. Ebstein, Assignment of transmitter powers by linear programming. IEEE Trans. Electromagn. Compat. **6**(2), 36-44 (1964)

55. D. Cai, T. Quek, C. Tan, S. Low, Max-min SINR coordinated multipoint downlink transmission-duality and algorithms. IEEE Trans. Signal Process. **60**(10), 5384–5395 (2012)
56. G. Caire, N. Jindal, M. Kobayashi, N. Ravindran, Multiuser MIMO achievable rates with downlink training and channel state feedback. IEEE Trans. Inf. Theory, **56**(6), 2845–2866 (2010)
57. G. Caire, S. Ramprashad, H. Papadopoulos, Rethinking network MIMO: cost of CSIT, performance analysis, and architecture comparisons, in *IEEE Transactions on Information Theory*, 2010
58. G. Caire, S. Shamai, On the achievable throughput of a multiantenna Gaussian broadcast channel, in *IEEE Transactions on Information Theory*, 2003
59. I. Csiszar, J. Korner, Broadcast channels with confidential messages, in *IEEE Transactions on Information Theory*, 1978
60. H. Dahrouj, W. Yu, Coordinated beamforming for the multicell multiantenna wireless system. IEEE Trans. Wirel. Commun. **9**(5), 1748–1759 (2010)
61. D. Dardari, V. Tralli, A. Vaccari, A theoretical characterization of nonlinear distortion effects in OFDM systems. IEEE Trans. Commun. **48**(10), 1755–1764 (2000)
62. K. Fan, Minimax theorems, in *Proceedings of National Academic Society*, 1953
63. C. Farsakh, J. Nossek, Channel allocation and downlink beamforming in an SDMA mobile radio system, in *IEEE International Symposium on Personal, Indoor and Mobile Radio Communications*, 1995
64. A. Abdelhadi, C. Clancy, Context-aware resource allocation in cellular networks (2014). arXiv preprint arXiv:1406.1910
65. Z. Kbah, A. Abdelhadi, Resource allocation in cellular systems for applications with random parameters, in *2016 International Conference on Computing, Networking and Communications (ICNC)* (IEEE, Piscataway, 2016), pp. 1–5
66. A. Abdel-Hadi, J. Michel, A. Gerstlauer, S. Vishwanath, Real-time optimization of video transmission in a network of AAVs, in *2011 IEEE Vehicular Technology Conference (VTC Fall)* (2011), pp. 1–5
67. M. Ghorbanzadeh, A. Abdelhadi, C. Clacy, *Cellular Communications Systems in Congested Environments Resource Allocation and End-to-End Quality of Service Solutions with MAT-LAB* (Springer, Berlin, 2017)
68. M. Ghorbanzadeh, Resource allocation and end-to-end quality of service for cellular communications systems in congested and contested environments. Ph.D. Thesis, Virginia Tech, 2015
69. M. Ghorbanzadeh, Y. Chen, K. Ma, C. Clancy, R. McGwier, A neural network approach to category validation of Android applications, in *IEEE Conference on Computing, Networking, and Communications (ICNC)*, 2013
70. A. Kumar, A. Abdelhadi, T.C. Clancy, A delay optimal MAC and packet scheduler for heterogeneous M2M uplink (2016). arXiv preprint arXiv:1606.06692
71. E. Hossain, Z. Han, H.V. Poor, *Smart Grid Communications and Networking* (Cambridge University Press, Cambridge, 2012)
72. A. Kumar, A. Abdelhadi, T.C. Clancy, An online delay-optimal iterative multiclass scheduler for generic M2M uplink, under submission
73. A. Gotsis, A. Lioumpas, A. Alexiou, Evolution of packet scheduling for machine-type communications over LTE: Algorithmic design and performance analysis, in *IEEE GLOBECOM Workshop* (2012), pp. 1620–1625
74. A. Kumar, A. Abdelhadi, T.C. Clancy, A delay-optimal packet scheduler for M2M uplink, in *IEEE MILCOM*, 2016
75. J.J. Nielsen, G.C. Madueño, N.K. Pratas, R.B. Sørensen, C. Stefanovic, P. Popovski, What can wireless cellular technologies do about the upcoming smart metering traffic? IEEE Commun. Mag. **53**, 41–47 (2015)
76. V. Annapureddy, V. Veeravalli, Sum capacity of MIMO interference channels in the low interference regime. IEEE Trans. Inf. Theory **57**(5), 2565–2581 (2011)

77. A. Abdelhadi, T. Clancy, An optimal resource allocation with frequency reuse in cellular networks (2015)
78. Y. Chen, M. Ghorbanzadeh, K. Ma, C. Clancy, R. McGwier, A hidden Markov model detection of malicious Android applications at runtime, in *2014 23rd Wireless and Optical Communication Conference (WOCC)*, 2014
79. M. Ghorbanzadeh, E. Visotsky, P. Moorut, W. Yang, C. Clancy, Radar inband and out-of-band interference into LTE macro and small cell uplinks in the 3.5 GHz band, in *2015 IEEE Wireless Communications and Networking Conference (WCNC)*, 2015
80. M. Ghorbanzadeh, E. Visotsky, P. Moorut, W. Yang, C. Clancy, Radar in-band interference effects on macrocell LTE uplink deployments in the U.S. 3.5 GHz band, in *2015 International Conference on Computing, Networking and Communications (ICNC)*, 2015
81. A. Abdel-Hadi, S. Vishwanath, On multicast interference alignment in multihop systems, in *IEEE Information Theory Workshop 2010 (ITW 2010)*, 2010
82. J. Jose, A. Abdel-Hadi, P. Gupta, S. Vishwanath, On the impact of mobility on multicast capacity of wireless networks, in *2010 Proceedings IEEE INFOCOM* (2010), pp. 1–5
83. M. Ghorbanzadeh, E. Visotsky, P. Moorut, W. Yang, C. Clancy, Radar interference into LTE base stations in the 3.5 GHz band, in *Phys. Commun.* **20**, 33–47 (2016)
84. H. Shajaiah, M. Ghorbanzadeh, A. Abdelhadi, C. Clancy, Application-aware resource allocation based on channel information for cellular networks, in *2019 IEEE Wireless Communications and Networking Conference (WCNC)* (2019), pp. 1–6
85. M. Ghorbanzadeh, A. Abdelhadi, C. Clancy, Application-aware resource allocation of hybrid traffic in cellular networks. IEEE Trans. Cogn. Commun. Netw. **3**(2), 226–241 (2017)
86. Y. Wang, A. Abdelhadi, A QoS-based power allocation for cellular users with different modulations (2015). abs/1507.07141
87. Y. Wang, A. Abdelhadi, Optimal power allocation for LTE users with different modulations (2015). abs/1507.07159
88. A. Abdelhadi, A. Khawar, T.C. Clancy, Optimal downlink power allocation in cellular networks (2014). abs/1405.6440
89. M. Li, Z. Chen, Y. Tan, A maxmin resource allocation approach for scalable video delivery over multiuser MIMO-OFDM systems, in *IEEE International Symposium on Circuits and Systems (ISCAS)*, 2011
90. A. Khawar, A. Abdelhadi, T.C. Clancy, Channel modeling between seaborne MIMO radar and MIMO cellular system (2015). abs/1504.04325
91. R.M.J.-W. Lee, N.B. Mandayam, A utility based approach for multi-class wireless systems. IEEE/ACM Trans. Netw **13**, 854–867 (2015)
92. S. Ahmed, J.S. Thompson, B. Mulgrew, Y. Petillot, Constant envelope waveform design for MIMO radar, in *International Conference on Acoustics, Speech, and Signal Processing* (2010), pp. 4066–4069
93. A. Khawar, A. Abdelhadi, T.C. Clancy, 3d channel modeling between seaborne MIMO radar and MIMO cellular system (2015). abs/1504.04333
94. P. Marbach, R. berry, Downlink power allocation and pricing for wireless networks. Proc. IEEE INFOCOM **3**, 1470–1479 (2002)
95. S. Ahmed, J.S. Thompson, Y.R. Petillot, B. Mulgrew, Finite alphabet constant-envelope waveform design for MIMO radar. IEEE Trans. Signal Process. **59**(11), 5326–5337 (2011)
96. A. Khawar, A. Abdelhadi, T.C. Clancy, Coexistence analysis between radar and cellular system in LoS channel (2015). abs/1506.07468
97. M. Andrews, K. Kumaran, K. Ramanan, A. Stolyar, P. Whiting, R. Vijayakumar, Providing quality of service over a shared wireless link. IEEE Commun. Mag. **39**(2), 150–154 (2001)
98. D. Fudenberg, J. Tirole, Nash Equilibrium: Multiple Nash Equilibria, Focal Points, and Pareto Optimality (MIT Press, Cambridge, 1991)
99. C. Shahriar, A. Abdelhadi, T.C. Clancy, Overlapped-MIMO radar waveform design for coexistence with communication systems (2015). abs/1502.04117

100. A. Khawar, A. Abdel-Hadi, T. Clancy, R. McGwier, Beampattern analysis for MIMO radar and telecommunication system coexistence, in *2014 International Conference on Computing, Networking and Communications (ICNC)* (2014), pp. 534–539
101. A. Khawar, A. Abdelhadi, T. Clancy, A mathematical analysis of cellular interference on the performance of s-band military radar systems, in *Wireless Telecommunications Symposium (WTS), 2014* (2014), pp. 1–8
102. A. Babaei, W.H. Tranter, T. Bose, A practical precoding approach for radar/communications spectrum sharing, in *Cognitive Radio Oriented Wireless Networks (CROWNCOM)* (2013), pp. 13–18
103. H. Kushner, P. Whiting, Convergence of proportional-fair sharing algorithms under general conditions. IEEE Trans. Wirel. Commun. **3**(4), 1250–1259 (2004)
104. Shajaiah, H., Khawar, A., Abdel-Hadi, A., Clancy, T.C., Resource allocation with carrier aggregation in LTE Advanced cellular system sharing spectrum with S-band radar, in *2014 IEEE International Symposium on Dynamic Spectrum Access Networks (DYSPAN)* (IEEE, Piscataway), pp. 34–37
105. A. Parekh, R. Gallager, A generalized processor sharing approach to flow control in integrated services networks: the single-node case. IEEE/ACM Trans. Netw. **1**(3), 344–357 (1993)
106. A. Demers, S. Keshav, S. Shenker, Analysis and simulation of a fair queuing algorithm. SIGCOMM Comput. Commun. Rev. **19**(4), 1–12 (1989)
107. A. Khawar, A. Abdelhadi, T.C. Clancy, On the impact of time-varying interference-channel on the spatial approach of spectrum sharing between S-band radar and communication system, in *Military Communications Conference (MILCOM)*, 2014

Chapter 3
Resource Allocation Without Channel

3.1 Introduction

In this chapter, we present a resource allocation framework based on proportional fairness (Sect. 3.2) that addresses some of the challenges mentioned in Chap. 1 such as optimal resource allocation, convexity for easy solution, UE prioritization, traffic change consideration, and traffic type awareness. The proposed resource allocation does not consider channel condition, the deterioration due to environment propagation loss effects. The items discussed in this chapter are as follows.

- Centralized and distributed architecture of resource allocation without channel modeling is provided.
- We show that the devised centralized architecture is a convex optimization.
- Solution algorithms for the centralized and distributed RF resource allocations are presented.
- Simulations to show the application of the proposed RF resource allocation in a cellular network are depicted.

The rest of this chapter proceeds as follows. Section 3.2 presents a centralized resource allocation for the utility proportional fairness resource allocation framework that is developed in this chapter by formulating the resource allocation in Sect. 3.3. Then, Sect. 3.4 proves mathematical equivalence of the centralized and distributed methods. Section 3.5 discusses the pros/cons of the two approaches. Section 3.6 compares the method developed here with some other methods in the literature. And, Sect. 3.7 summarizes this chapter.

M. Ghorbanzadeh, A. Abdelhadi, *Practical Channel-Aware Resource Allocation*,
https://doi.org/10.1007/978-3-030-73632-3_3

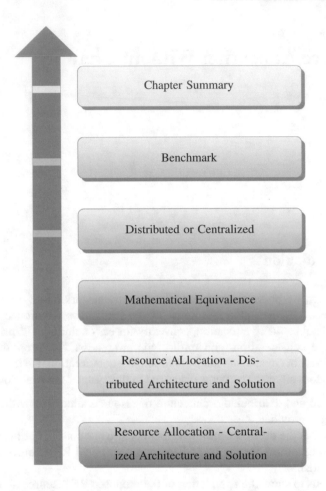

3.2 Resource Allocation: Centralized Architecture and Solution

Mobile broadband services have observed an unceasing demand for more radio resources during recent years owing to the dramatic increase of mobile broadband subscribers, outgrowth of the mobile broadband generated traffic volume [1], migration of cellular communication providers from offering a single service to a multi-service framework such as multimedia telephony and mobile TV [2], emergence of technology line M2MCs and Internet-of-things (IoT) [3–6], and prevalence of smartphones concurrently running delay-tolerant and real-time applications with distinctive QoS requirements [7, 8] create an urgency to dynamically provide various bit rates to applications in order to elevate users' QoE that directly speaks to subscriber churn rate [2]; hence, creating a service differentiation mechanism as part of the resource allocation is important. Besides, application temporal usage changes directly impact the amount of generated traffic, and therefore, inclusion of application differentiation in resource allocation is another factor. Moreover, MNO's ability to adopt a resource allocation with user differentiation [2] (corporate users vs. private users, post-paid users vs. pre-paid users, and privileged users vs. roaming users) further fine-tunes resource allocation. Consequently, resource allocation methods adopting such enhancements can accommodate diverse hybrid traffic by accounting for all the aforesaid matters. Even so, a majority of resource allocation proposals fizzle to collectively consider the formerly mentioned concerns. The proposed allocation method in this chapter uses Levenberg–Marquardt method [9–11], channel quality indicator (CQI) [7, 12–18], and adaptive modulation and coding schemes (MCSs) [19–21]. The proposed method is a convex utility proportional fairness maximization formulation for an optimal resource allocation and outfitted with application status, user, and service differentiation parameterized application weights, UE weights, and application utility functions. The weights provided by network providers assign a foreground-running application a higher application status weight than those of the background-running ones. Mobile subscribers concurrently run multiple applications whose utility functions and state depend on the traffic type and temporal usage percentage. Putting traffic differentiation under a utility proportional fairness to prioritize real-time applications over delay-tolerant ones in order to fulfill QoS requirements is important. Then, the formalized resource allocation is solved analytically.

The objective of the resource allocation is to determine optimal rates that would be allocated to the UE applications such that real-time applications are given priority over delay-tolerant ones, no user ever is dropped (defined by a zero rate allocation), application usage changes are considered, and subscription-based treatments are considered. It is assumed that each UE includes concurrently running real-time and delay-tolerant applications, represented by sigmoidal and logarithmic application utility functions mathematically as was explained in Sect. 2.3. A cellular network single cell with a BS serving M UEs (here $M = 6$ depicted in Fig. 3.1) is considered such that a UE simultaneously runs real-time and delay-tolerant applications shown by the sigmoidal and logarithmic utility functions in Sect. 2.3. The rate assigned by the BS to the ith UE is referred to as r_i. An aggregate utility function is denoted as $V_i(r_i)$, which relates to the individual UE application utility function according to Eq. (3.1) in which r_{ij}, $U_{ij}(r_{ij})$, and α_{ij} are, respectively, the rate allocated to, application utility function of, and application usage percentage of the jth application of the ith UE. Since one can write $\sum_{j=1}^{N_i} \alpha_{ij} = 1$ and $r_i = \sum_{j=1}^{N_i} r_{ij}$ in which N_i is the number of running applications on the ith UE and r_i is the resource assigned by the BS, the former states the addition of the ith UE application usage percentages equals a 100% usage percentage, and the letter implies the ith UE rate is the addition of its N_i application resource assignments. Before diving deeper, the following lemma is presented and is used in later in this chapter.

$$V_i(r_i) = \prod_{j=1}^{N_i} U_{ij}^{\alpha_{ij}}(r_{ij}) \tag{3.1}$$

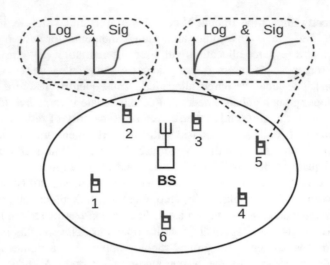

Fig. 3.1 System model: single cell, within the cellular network, with a BS covering $M = 6$ UEs each with simultaneously running delay-tolerant and relay-time applications represented by logarithmic and sigmoidal utility functions, respectively

Lemma 3.2.1 *Natural logarithm of the aggregate utility function, i.e.* $\log(V_i(r_i))$, *is a strictly concave function.*

Proof From Eq. (3.1), we can write $\log V_i(r_i) = \sum_{j=1}^{N_i} \alpha_{ij} \log U_{ij}(r_{ij})$, where $U_{ij}(r_{ij}) > 0$ in accordance with Chap. 2 utility function properties. Also, logarithmic utilities concavity stems out $U'_{ij}(r_{ij}) = \frac{dU_{ij}(r_{ij})}{dr_{ij}} > 0$ and $U''_i(r_{ij}) = \frac{d^2 U_{ij}(r_{ij})}{dr_{ij}^2} < 0$, resulting in $\frac{d \log(U_{ij}(r_{ij}))}{dr_{ij}} = \frac{U'_{ij}(r_{ij})}{U_{ij}(r_{ij})} > 0$ due to $U_{ij}(r_{ij}) > 0$ and $U'_{ij}(r_{ij}) > 0$ and in $\frac{d^2 \log(U_{ij}(r_{ij}))}{dr_{ij}^2} = \frac{U''_{ij}(r_{ij})U_{ij}(r_{ij}) - U'^2_{ij}(r_{ij})}{U^2_{ij}(r_{ij})} < 0$ due to $U''_{ij}(r_{ij}) < 0$. Thus, the logarithmic utility natural logarithm is strictly concave. On the flip side, for a sigmoidal utility $U_{ij}(r_{ij})$ with $0 < r_{ij} < R$, we have the following inequalities among which the first owes to the sigmoidal function's continuity and $0 \le U_{ij}(r_{ij}) < 1$ and the rest are utter algebraic manipulation of the first one.

$$0 < c_{ij}\left(\frac{1}{1 + e^{-a_{ij}(r_{ij} - b_{ij})}} - d_{ij}\right) < 1 \quad \wedge \quad d_{ij} < \frac{1}{1 + e^{-a_{ij}(r_{ij} - b_{ij})}} < \frac{1 + c_{ij}d_{ij}}{c_{ij}} \quad \wedge$$

$$\frac{1}{d_{ij}} > 1 + e^{-a_{ij}(r_{ij} - b_{ij})} > \frac{c_{ij}}{1 + c_{ij}d_{ij}} \quad \wedge \quad 0 < 1 - d_{ij}(1 + e^{-a_{ij}(r_{ij} - b_{ij})}) < \frac{1}{1 + c_{ij}d_{ij}}$$

In the aforesaid inequalities, "$\wedge$" indicates logical "AND" which is "true" of its both component statements are "true." For $0 < r_{ij} < R$, we have the following inequalities, of which the first results from first additive's denominator positivity in addition to the formerly derived statement $0 < 1 - d_{ij}(1 + e^{-a_{ij}(r_{ij}-b_{ij})}) < \frac{1}{1+c_{ij}d_{ij}}$ as well as other constituents' positivity and the last one is verifiable by investigating its terms algebraically. Hence, the sigmoidal utility natural logarithm is strictly concave. As such, the application utility functions $U_{ij}(r_{ij}) > 0$ of the system model (Eq. (3.1)) have strictly concave natural logarithms, meaning that the aggregated utility $\log V_i(r_i) = \sum_{j=1}^{N_i} \alpha_{ij} \log U_{ij}(r_{ij})$ is strictly concave.

$$\frac{d}{dr_{ij}} \log U_{ij}(r_{ij}) = \frac{a_{ij}d_{ij}e^{-a_{ij}(r_{ij}-b_{ij})}}{1 - d_{ij}(1 + e^{-a_{ij}(r_{ij}-b_{ij})})} + \frac{a_{ij}e^{-a_{ij}(r_{ij}-b_{ij})}}{(1 + e^{-a_{ij}(r_{ij}-b_{ij})})} > 0$$

$$\frac{d^2}{dr_{ij}^2} \log U_{ij}(r_{ij}) = \frac{-a_{ij}^2 d_{ij}e^{-a_{ij}(r_{ij}-b_{ij})}}{c_{ij}\left(1 - d_{ij}(1 + e^{-a(r_{ij}-b_{ij})})\right)^2} + \frac{-a_{ij}^2 e^{-a_{ij}(r_{ij}-b_{ij})}}{(1 + e^{-a_{ij}(r_{ij}-b_{ij})})^2} < 0$$

$$(3.2)$$

Lemma 3.2.1 proves the concavity of the aggregate utility function natural logarithm that itself is the utility function used in the resource allocation presented in Sect. 3.2. Next, Lemma 3.2.2 presents another useful result.

Lemma 3.2.2 *The aggregate utility function $V_i(r_i)$ and the slope curvature function $\frac{\partial \log V_i(r_i)}{\partial r_i}$ have its inflection point at $r_i = r_{ij}^s \approx r_{ij}^{inf}$ for jth application utility function U_{1j} and are convex for $r_{ij} > \max_j r_{ij}^s$.*

Proof For the ith UE aggregate utility $V_i(r_i)$, let $S_i(r_i) = \frac{\partial \log V_i(r_i)}{\partial r_i}$ be the aggregate utility slope curvature function, let $S_{ij}(r_{ij}) = \frac{\partial \log U_{ij}(r_{ij})}{\partial r_{ij}}$ be the jth application utility slope curvature function, and let N_i^S be the number of sigmoidal utility functions. Taking the logarithm and derivative of both sides of Eq. (3.1) leads to Eq. (3.3).

$$S_i(r_i) = \frac{\partial \log V_i(r_i)}{\partial r_i} = \frac{\partial}{\partial r_i} \sum_{j=1}^{N_i} \alpha_{ij} \log U_{ij}(r_{ij}) = \sum_{j=1}^{N_i} \alpha_{ij} \frac{\partial \log U_{ij}(r_{ij})}{\partial r_{ij}}$$

$$= \sum_{j=1}^{N_i} \alpha_{ij} S_{ij}(r_{ij}) = \sum_{j=1}^{N_i^S} \alpha_{ij} S_{ij}(r_{ij}) + \sum_{j=N_i^S+1}^{N_i} \alpha_{ij} S_{ij}(r_{ij}) \qquad (3.3)$$

Taking the 1st and 2nd derivatives of Eq. (3.3) yields Eq. (3.4) and (3.5).

$$\frac{\partial S_i}{\partial r_i} = \sum_{j=1}^{N_i^S} \left\{ \frac{-\alpha_{ij}a_{ij}^2 d_{ij}e^{-a_{ij}(r_{ij}-b_{ij})}}{c_{ij}\left(1 - d_{ij}(1 + e^{-a_{ij}(r_{ij}-b_{ij})})\right)^2} + \frac{\alpha_{ij}a_{ij}^2 e^{-a_{ij}(r_{ij}-b_{ij})}}{\left(1 + e^{-a_{ij}(r_{ij}-b_{ij})}\right)^2} \right\}$$

$$-\sum_{j=N_i^S+1}^{N_i}\left\{\frac{\alpha_{ij}k_{ij}^2}{(1+k_{ij}r^{max})\log(1+k_{ij}r_{ij})^2}\right\} \tag{3.4}$$

It is easy to demonstrate that $\forall\, r_i$, $\frac{\partial S_i}{\partial r_i}<0$. Denoting the 1th term of Eq. (3.4) and the 2nd and 3rd terms of Eq. (3.5) as, respectively, S_i^1, S_i^2, and S_i^3, Eq. (3.2) for which the properties in Eq. (3.2) are considerable is concluded.

$$\frac{\partial^2 S_i}{\partial r_i^2}=\sum_{j=1}^{N_i^S}\left\{\frac{d_{ij}e^{-a_{ij}(r_{ij}-b_{ij})}(1-d_{ij}(1-e^{-a_{ij}(r_{ij}-b_{ij})}))}{c_{ij}\left(1-d_{ij}(1+e^{-a_{ij}(r_{ij}-b_{ij})})\right)^3}\right.$$
$$\left.+\frac{e^{-a_{ij}(r_{ij}-b_{ij})}(1-e^{-a_{ij}(r_{ij}-b_{ij})})}{\left(1+e^{-a_{ij}(r_{ij}-b_{ij})}\right)^3}\right\}\times a_{ij}^3\alpha_{ij}$$
$$-\sum_{j=N_i^S+1}^{N_i}\left\{\frac{\alpha_{ij}k_{ij}^2(\log(1+k_{ij}r_{ij})-1)}{(1+k_{ij}r_{ij})^2\log^2(1+k_{ij}r_{ij})}\right\} \tag{3.5}$$

Equation (3.2) shows that the slope curvature function S_i has an inflection point $r_i=r_{ij}^s\approx b_{ij}=r_{ij}^{\inf}$ and it changes from a convex function in the vicinity of the origin to a concave function before the inflection point at $r_{ij}=r_{ij}^s$ to a convex function after the inflection point.

$$\begin{cases}S_i^1=\frac{\alpha_{ij}a_{ij}^3e^{a_{ij}b_{ij}}(e^{a_{ij}b_{ij}}+e^{-a_{ij}(r_{ij}-b_{ij})})}{(e^{a_{ij}b_{ij}}-e^{-a_{ij}(r_{ij}-b_{ij})})^3}\\[2mm] S_i^2=\frac{a_{ij}^3\alpha_{ij}e^{-a_{ij}(r_{ij}-b_{ij})}(1-e^{-a_{ij}(r_{ij}-b_{ij})})}{\left(1+e^{-a_{ij}(r_{ij}-b_{ij})}\right)^3}\\[2mm] S_i^3=\frac{\alpha_{ij}k_{ij}^2(\log(1+k_{ij}r_{ij})-1)}{(1+k_{ij}r_{ij})^2\log^2(1+k_{ij}r_{ij})}\end{cases} \tag{3.6}\qquad \begin{cases}\lim_{r_i\to 0}S_i^1=\infty,\\ \lim_{r_i\to b_{ij}}S_i^1=0\ \text{for}\ b_{ij}\gg\frac{1}{a_{ij}}\ \forall\ j\\ S_i^2(b_{ij})=0\\ S_i^2(r_{ij}>b_{ij})>0\\ S_i^2(r_{ij}<b_{ij})<0\\ S_i^3(r_{ij}>0)>0\end{cases} \tag{3.7}$$

Next, Lemma 3.2.3 used later in this chapter proves that the logarithms of aggregate and application utility functions are invertible functions and the inverse functions are strictly decreasing.

Lemma 3.2.3 *The aggregate and application utility slope curvature functions* $S_i(r_i)=\frac{\partial\log V_i(r_i)}{r_i}$ *and* $S_{ij}(r_{ij})=\frac{\partial\log U_{ij}(r_{ij})}{r_{ij}}$ *are invertible and their inverse functions* $r_i=S_i^{-1}(.)$ *and* $r_{ij}=S_{ij}^{-1}(.)$ *are strictly decreasing.*

Proof The concavity of the logarithmic application utility function U_{ij} stems out $U_{ij}'(r_{ij})=\frac{\partial U_{ij}(r_{ij})}{\partial r_{ij}}>0$ and $U_{ij}''(r_{ij})=\frac{\partial^2 U_{ij}(r_{ij})}{\partial r_{ij}^2}<0$, and Lemma 3.2.1 yields

$S_{ij}(r_{ij}) = \frac{\partial \log(U_{ij}(r_{ij}))}{\partial r_{ij}} = \frac{U'_{ij}(r_{ij})}{U_{ij}(r_{ij})} > 0$ and $\frac{\partial S_{ij}(r_{ij})}{\partial r_{ij}} = \frac{U''_{ij}(r_{ij})U_{ij}(r_{ij}) - U'^2_{ij}(r_{ij})}{U^2_{ij}(r_{ij})} <$ 0. Also, for the application utility function, we have $U_{ij}(r_{ij}) > 0$, $U_{ij}(r_{ij})$ is increasing, and it is twice differentiable with respect to r_{ij}. So, $S_{ij}(r_{ij})$ of the logarithmic application utility function is a strictly decreasing function. From Eq. (3.2), for the sigmoidal application utility function $U_{ij}(r_{ij})$ where $0 < r_{ij} < R$, inequality (3.8) can be written and show that $S_{ij}(r_{ij})$ of the sigmoidal application utility function is a strictly decreasing function.

$$S_{ij}(r_{ij}) > 0, \frac{\partial}{\partial r_{ij}} S_{ij}(r_{ij}) < 0 \tag{3.8}$$

Combining Eq. (3.3) and inequality (3.8) stems out inequality (3.9). Thus, $S_{ij}(r_{ij})$ and $S_i(r_i)$ of all application utility functions in Chap. 2 are strictly decreasing functions that result in the slope curvature functions $S_{ij}(r_{ij})$ and $S_i(r_l)$ being invertible and the inverse functions being strictly decreasing.

$$S_i(r_i) = \sum_{j=1}^{N_i^S} \alpha_{ij} S_{ij}(r_{ij}) + \sum_{j=N_i^S+1}^{N_i} \alpha_{ij} S_{ij}(r_{ij}) > 0$$

$$\frac{\partial S_i(r_i)}{\partial r_i} = \sum_{j=1}^{N_i^S} \alpha_{ij} \frac{\partial S_{ij}(r_{ij})}{\partial r_{ij}} + \sum_{j=N_i^S+1}^{N_i} \alpha_{ij} \frac{\partial S_{ij}(r_{ij})}{\partial r_{ij}} < 0 \tag{3.9}$$

Next, we develop the formulation to assign resources to the applications of UEs with the aggregated utility as in Eq. (3.1). We consider the resource allocation as an optimization that assigns resources directly by the BS in one stage as in Eq. (3.10). Since the assignment occurs in one stage, this method is conducive to being implemented in a centralized architecture.

$$\max_{\mathbf{r}} \quad \prod_{i=1}^{M} \left(\prod_{j=1}^{N_i} U_{ij}^{\alpha_{ij}}(r_{ij}) \right)^{\beta_i}$$

$$\text{subject to} \quad \sum_{i=1}^{M} \sum_{j=1}^{N_i} r_{ij} \le R, r_{ij} \ge 0, \quad i = 1, 2, \ldots, M, \quad j = 1, 2, \ldots, N_i \tag{3.10}$$

Here, for M UEs covered by the BS, $\mathbf{r} = [r_1, r_2, \ldots, r_M]$ is the UE allocated rate vector, R is the resources available at the BS, and β_i is a subscription weight of the ith UE. The centralized architecture's objective function $\prod_{i=1}^{M} \left(\prod_{j=1}^{N_i} U_{ij}^{\alpha_{ij}}(r_{ij}) \right)^{\beta_i}$ mathematically corresponds to the statement (3.11). Then, reformulating Eq. (3.10) as Eq. (3.12), referred to hereinafter as the log-centralized problem, Corollary 3.2.4 is proved and will be used later.

$$\sum_{i=1}^{M} \beta_i \sum_{j=1}^{N_i} \alpha_{ij} \log U_{ij}(r_{ij}) \tag{3.11}$$

$$\max_{\mathbf{r}} \quad \sum_{i=1}^{M} \beta_i \sum_{j=1}^{N_i} \alpha_{ij} \log U_{ij}(r_{ij})$$

$$\text{subject to} \quad \sum_{i=1}^{M}\sum_{j=1}^{N_i} r_{ij} \leq R, r_{ij} \geq 0, \quad i = 1, 2, \ldots, M, \ j = 1, 2, \ldots, N_i \tag{3.12}$$

Corollary 3.2.4 substantiates the existence of a globally optimal solution for the resource allocation optimization in Eq. (3.10).

Corollary 3.2.4 *The resource allocation optimization in Eq. (3.10) is convex and has a unique tractable globally optimal solution.*

Proof Substantiating Lemma 3.2.1 yields the application utility natural logarithm concavity that in turn stems out the convexity of the log-centralized optimization in Eq. (3.12) [22, 23]. This results in the convexity of the equivalent resource allocation optimization in Eq. (3.10) and proves the existence of a tractable globally optimal solution [22].

Similarly to [24, 25], the duality for convex optimization problem in Eq. (3.10) solves it efficiently with UE and BS segments of the solution shown in, respectively, Algorithms 2 and 3 whose execution (Fig. 3.2) starts by UEs transmitting their application utility parameters to the BS that solves the entire optimization. The rates are values r_{ij} solving equation $\frac{\partial \log U_{ij}(r_{ij})}{\partial r_{ij}} = p(n)$ and are geometrically the intersection of the time varying shadow price, horizontal line $y = p(n)$, with the curve $y = \frac{\partial \log U_{ij}(r_{ij})}{\partial r_{ij}}$. Next, Sect. 3.2.1 provides simulations for the centralized architecture developed in this chapter.

Algorithm 2 UE centralized algorithm

---
loop
  UEs send application utility parameters $\{a_{ij}, b_{ij}, \alpha_{ij}, k_{ij}, r_{ij}^{\max}\}$ to BS.
  UEs receive rates $r_i^{\text{opt}} = \{r_{i1}^{\text{opt}}, r_{i2}^{\text{opt}}, \ldots, r_{iN_i}^{\text{opt}}\}$ from BS.
  UEs assign resources r_{ij}^{opt} internally to jth applications.
end loop

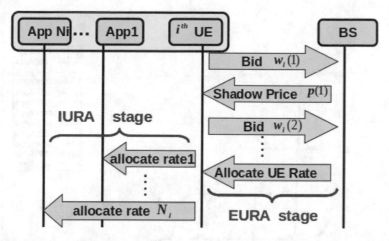

Fig. 3.2 Centralized algorithm assigns resources to the UE applications in a single stage where UEs send their application utility parameters to their serving BS that calculates the optimal application rates and sends them to the UEs

---

Algorithm 3 BS centralized algorithm

---

loop

  BS receives application utility parameters $\{a_{ij}, b_{ij}, \alpha_{ij}, k_{ij}, r_{ij}^{\max}\}$ from its served UEs.

  BS solves $\mathbf{r} = \arg\max_{\mathbf{r}} \sum_{i=1}^{M} \beta_i \sum_{j=1}^{N_i} \alpha_{ij} \log U_{ij}(r_{ij}) - p(\sum_{i=1}^{M} \sum_{j=1}^{N_i} r_{ij} - R)$. {where $\mathbf{r} = \{r_1, r_2, \ldots, r_M\}$ and $r_i = \{r_{i1}, r_{i2}, \ldots, r_{iN_i}\}$}

  BS sends resources $r_i = \{r_{i1}, r_{i2}, \ldots, r_{iN_i}\}$ to ith UE.

end loop

3.2.1   Centralized Architecture Simulation

A cell with $M = 6$ UEs and 1 BS, depicted in Fig. 3.1, with each UE concurrently running a delay-tolerant and a real-time application modeled by logarithmic and sigmoidal application utility functions in Table 2.1 is considered. The sigmoidal utility parameters $a = 5$ and $b = 10$ approximate a step function at rate $r = 5$, parameters $a = 3$ and $b = 15$ are an approximation of a real-time application with an inflection point at rate $r = 15$, parameters $a = 1$ and $b = 25$ are estimations of real-time applications with the inflection point at $r = 25$, and logarithmic utilities with $r^{\max} = 100$ and distinct k_i parameters represent delay-tolerant applications. The plot of application utility functions is shown in Fig. 2.1, which reaffirms the properties in Chap. 2, i.e. increasing utility functions and zero valued at zero rates. Besides, the 1st derivative of the utility function natural logarithm, denoted as $S_{ij}(r_{ij})$, is shown in Fig. 3.3a depicting $S_{ij}(r_{ij})$ positivity and decreasingness affirming Lemma 3.2.3. Then, the resource

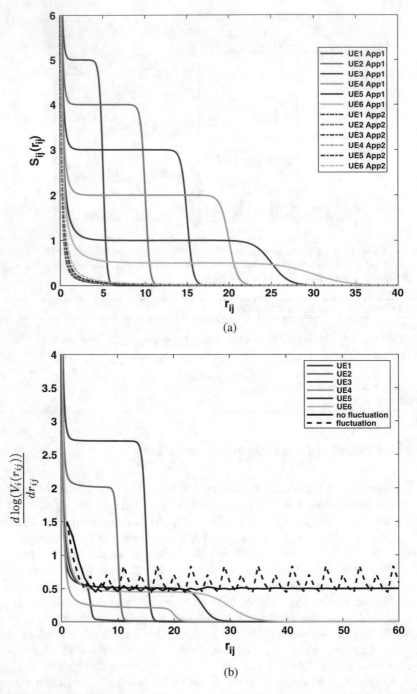

(a)

(b)

Fig. 3.3 Utility slope curvature functions $S_{ij}(r_{ij})$ are illustrated in (**a**) where identical colors are for one UE application. (**b**) illustrates the aggregate slope curvature functions $S_i(r_i)$. The lack of decay functions stems out the system instability shown in the shadow price oscillation. (**a**) Application slope curvature functions vs. application rates. (**b**) Aggregate slope curvature functions vs. application rates

allocations in Algorithms 2 and 3 are applied. The application status weight vector in Eq. (3.1) is set to $\boldsymbol{\alpha} = \{\alpha_{11}, \alpha_{21}, \alpha_{31}, \alpha_{41}, \alpha_{51}, \alpha_{61}, \alpha_{12}, \alpha_{22}, \alpha_{32}, \alpha_{42}, \alpha_{52}, \alpha_{62}\}$, where α_{ij} is the jth application status weight of the ith UE. It is noteworthy that $\alpha_{i1} + \alpha_{i2} = 1$ as expected. Figure 3.3b shows UE slope curvatures. These affirm Lemma 3.2.2 which says that $S_i(r_i)$ inflection points are at the application utility function's inflection points and also confirm Lemma 3.2.3 which says that $S_i(r_i)$ functions are strictly decreasing. The termination threshold is set to $\delta = 10^{-4}$, the BS resources to $R = 10, 15, 20, \ldots, 200$, and application status weights to $\boldsymbol{\alpha} = \{0.1, 0.5, 0.9, 0.1, 0.5, 0.9, 0.9, 0.5, 0.1, 0.9, 0.5, 0.1\}$. Applying Algorithms 2 and 3 produces application rates vs. R in Fig. 3.4a; initially, the UEs are allocated some rates since they have real-time applications requiring immediate rate allocations, increasing R increases the UE rates, while smaller R causes UEs with high QoS applications to bid higher to obtain resources. For example, UE2 has a real-time application so bids initially higher to get a fast allocation as shown in Fig. 3.4a. Then, the resource allocation algorithm assigns rates as portrayed in Fig. 3.4b depicting application rates $\{r_{ij}|i \in \{1, \ldots, 6\} \wedge j \in \{1, 2\}\}$ vs. R. First, resources are allocated to the real-time applications as they have stringent QoS requirements. When R exceeds the sigmoidal utilities' inflection points sum $\sum b_{ij}$, BS can assign more resources to the logarithmic utility applications; this is revealed in the rate/bid increase when $R > \sum b_{ij} = 105$ in Fig. 3.4a.

3.2.2   Centralized Resource Allocation in Real-World Implementation

This section shows that implementing the centralized resource allocation in a real-world network on a router improves the users' QoE by eliminating real-time traffic buffering and reducing the network's total resource consumption and operation expenditure (OPEX). The implementation architecture is shown in Fig. 3.5a where 2 UEs connect through a Wi-Fi access point (AP) to the Internet and the resource allocation is implemented on a resource broker (RB) logical entity installed on a router shaping the UE traffic received by the AP by the resource allocation scheme. To implement the scenario in Fig. 3.5a on a real-world network, we leverage a personal computer (PC) to configure the network in a distributed manner to decrease the processing load through a virtual machine (VM) architecture [26] in Fig. 3.5b. Here, we have employed a single-socket IBM x3250 M4 server [27] with 2 physical and 2 peripheral component interconnect (PCI)-enabled ports [28] to create 2 three-interfaced VMs. One of the VMs hosts the RB entity and another forms an enforcement engine to manage rate assignments via an onboard router traffic control, and another VM is a dedicated file server. The 2 VMs are annotated as "Guest 1" and "Guest 2" in Fig. 3.5b where the router and RB are, respectively, on "Guest 1" and "Guest 2." Additionally, we create 3 virtual switches intended for the phone network, for Internet-connected external devices, and for network

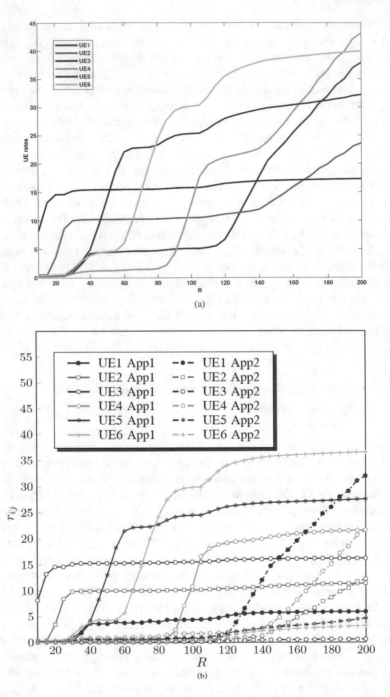

Fig. 3.4 (**a, b**) shows optimal UE and application rates vs. R where UE applications are identically colored. The real-time applications get more resources initially than do the delay-tolerant ones due to the formers' stringent QoS. (**a**) UE assigned rates. (**b**) Application assigned rates

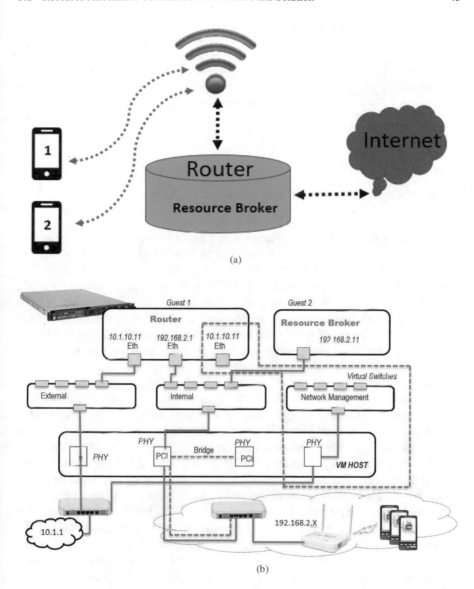

Fig. 3.5 UEs connect to Internet via Wi-Fi and run 1 delay-tolerant and 1 real-time application. RB entity assigns resources as it includes the resource allocation algorithm. Real-world network architecture has 2 VMs "Guest 1" and "Guest 2" that host the router and resource allocation, respectively. (**a**) Implementation of high level diagram. (**b**) Implementation details

Table 3.1 Network throughput in Fig. 3.5b without total bandwidth constraints and resource allocation application

Scenario	Phone 1	Phone 2	Phone 3	Rate (Mbps)
1	Video streaming 1	Video streaming 2	–	3.492
2	Video streaming 1	HTTP 1	–	4.050
3	Video streaming 1	–	–	0.951
4	Video streaming 2	—	–	2.11
5	HTTP 1	HTTP 1	–	4.262
6	HTTP 1	HTTP 2	–	24.866
7	HTTP 1	HTTP 1	Video streaming 1	13.747

Network throughput without any application and constraint is 32 Mbps

maintenance issues. The smart phones and their APs are on their own private Ethernet network [29] with IP address 192.168.2.1 [30–35]. The router gateway settings enable connections to the office network (Intranet) and the Internet. The smart phones run video streaming and HTTP download applications. The traffic generated by the real-time/delay-tolerant video streaming/HTTP applications is delay-tolerant/real-time, and we apply the resource allocation to obtain application rates. QoE adverse impact is reflected in the video streaming buffering and HTTP traffic incomplete downloads, while the lack thereof shows an acceptable QoE. In Fig. 3.5b, the small number of phones in a high throughput Wi-Fi network is concerning; to show the traffic shaping effects of the resource allocation, we restrict the overall network bandwidth to 1 mega bits per second (Mbps). The video streaming and HTTP applications throughput without network constraint and resource allocation algorithm application is shown in Table 3.1. This table shows that the low rate video streaming 1 and high rate video streaming 2 in addition to a small size download HTTP 1 and large download HTTP 2 run on the three UEs. The network speed with the absence of any applications is measured at 32 Mbps; for example, the 1st scenario shows that the 2 phones run video streaming 1, the bandwidth consumption is 3.492 Mbps, and the single UE 3rd scenario observes a bandwidth consumption 0.951 Mbps. Nonetheless, the bandwidth consumption increases to 2.11 Mbps due to video streaming 2 in the 4th scenario, and a similar increase is observed during the transition from the 5th scenario to the 6th one where the high rate HTTP 2 replaces the low rate HTTP 1 at the 2nd phone producing 24.866 Mbps. The final scenario observes a lower rate 13.747 Mbps vs. the 6th scenario (24.866 Mbps) despite adding a video streaming 1 application to phone 3 in the last case. This can be explained by the need for more data transfer over a longer time period as opposed to the 6th case.

Table 3.2 Smart phone applications under $R = 1$ Mbps bandwidth constraint with/without resource allocation shaping the traffic .

Phone	IP	Traffic type	Application
UE 1	192.168.2.57	Inelastic	Video streaming 1
UE 2	192.168.2.98	Elastic	HTTP 1

Next, the overall network bandwidth is contained to $R = 1$ Mbps, and only 2 phones in Fig. 3.5b running video streaming 1 and HTTP 1 as in Table 3.2 are placed in the network. Without the resource allocation, video streaming 1 incurs multiple buffering, the total average bandwidth is 0.963 Mbps, and HTTP 1 download completed in 1200 s. Using Wireshark [36, 37], Fig. 3.6a traces are observed, and both applications sharing a 1 Mbps throughput are illustrated on the black curve with bursty transmission intervals. At times, HTTP 1 application (the red curve) uses up the available resources shown as the red curve reaching the black curve and zeroing the video streaming 1 throughput depicted by the green curve reaching the time axis. The same scenario in Fig. 3.5b is repeated with the resource allocation where UE 1 and UE 2 register with the RB entity that computes the rates enforced at the router. The throughputs for the video streaming 1 and HTTP 1 are 731 and 267 kbps, respectively, the algorithm converges in 528 milliseconds (ms), and the throughput is 0.758 Mbps. The rates of the video streaming 1 and HTTP 1 with the overall $R = 1$ Mbps constraint are shown in Fig. 3.6b, where the black curve refers to the available resources. Video streaming 1 (the green curve) consumes more resources than the HTTP 1 application that downloads in 2650 s (Table 3.3), and there are instances where video streaming 1 throughput becomes 0 during which HTTP 1 has more resources, no video streaming buffering occurs, and the video watching experience gets better compared to the unshaped traffic situation in Fig. 3.6a. This real-world implementation shows that the resource allocation elevates the QoE and uses less resource that decreases the OPEX. For example, the bandwidth consumption went from 0.963 Mbps to 0.758 Mbps.

3.2.3 MATLAB Code for Centralized Resource Allocation

The following MATLAB code plots the application utility functions given in Chap. 2. Lines 1–12 initialize the application utility parameters in Fig. 2.1 using Table 2.1 and initialize the application slope curvature functions in Fig. 3.3a.

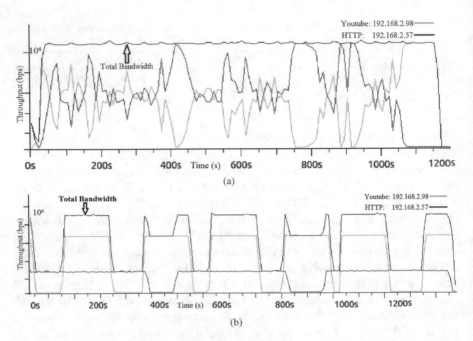

Fig. 3.6 Wireshark throughput analysis without the resource allocation in Fig. 3.5b and with parameters in Table 3.2 under a bandwidth constraint $R = 1$ Mbps: Black/Green/Red curve shows the network/video streaming 1/HTTP 1 application throughput, HTTP 1 downloads in 1200 s, video streaming 1 shows buffering as HTTP 1 (the red curve) uses up the entire resources. Wireshark throughput analysis with the resource allocation in Fig. 3.5b with parameters in Table 3.2 under a bandwidth constraint $R = 1$ Mbps: Black/Green/Red curve shows the network/video streaming 1/HTTP 1 throughput, HTTP 1 downloads in 2650 s, and no video streaming 1 buffering occurs. (**a**) Rates with no shaping. (**b**) Rates with shaping

```
1   syms x
2   alpha = [0.1 0.5 0.9 0.1 0.5 0.9; 0.9 0.5 0.1 0.9 0.5 0.1];
3   k = [15 12 9 6 3 1];
4   a = [5 4 3 2 1 0.5];
5   b = [5 10 15 20 25 30];
6   c = (1 + exp(a .* b)) ./ (exp(a .* b));
7   d = 1 ./ (1 + exp(a .* b));
8
9   for i = 1: length(a)
10      % Sigmoid utility function
11      y(i) = c(i).*(1./(1+exp(-a(i).*(x-b(i))))-d(i));
12      % Logarithmic utility function
13      y2(i) = log(k(i).*x+1)./(log(k(i).*100+1));
14      m(i) = exp(-a(i).*(x-b(i)));
15      % Diff Sigmoid utility function
16      dy_sig(i) = a(i).*m(i)./((1+m(i)).*(1-d(i).*(1+m(i))));
17      % Diff Logarithmic utility function
18      dy_log(i) = k(i)./((1+k(i).*x).*log(1+k(i).*x));
```

```
19  end
20  %%%%%%%%% the log utility functions %%%%%%%%%%
21  z = log(y);
22  z2 = log(y2);
23  %%%%%%%%%% the multiple utility functions %%%%%
24  u = alpha(1,:).*z + alpha(2,:).*z2;
25  u = z + 10*z2;
26  v = alpha(1,:).*y + alpha(2,:).*y2;
27  % Diff aggregated utility function
28  y_mul = alpha(1,:).*dy_sig + alpha(2,:).*dy_sig;
29  y_mul2 = alpha(1,:).*dy_sig ;
30  y_mul3 = alpha(2,:).*dy_sig;
31  y_mul = dy_sig + 10*dy_log;
32  y_mul2 = dy_sig ;
33  y_mul3 = 10*dy_log;
34  for i = 1: length(a)
35      for j = 2:1: 1000
36          x0(j) = 0.1 * j ;
37          yy(j,i) = subs(y(i),x0(j));
38          yy2(j,i) = subs(y2(i),x0(j));
39          dy(j,i) = diff(y(i),x);
40          dy2(j,i) = diff(y2(i),x);
41          dyy(j,i) = subs(dy(j,i),x,x0(j));
42          dyy2(j,i) = subs(dy2(j,i),x,x0(j));
43          zz(j,i) = subs(z(i),x0(j));
44          zz2(j,i) = subs(z2(i),x0(j));
45          dz(j,i) = diff(z(i),x);
46          dz2(j,i) = diff(z2(i),x);
47          dzz(j,i) = subs(dz(j,i),x,x0(j));
48          dzz2(j,i) = subs(dz2(j,i),x,x0(j));
49          uu(j,i) = subs(u(i),x,x0(j));
50          du(j,i) = diff(u(i),x);
51          duu(j,i) = subs(du(j,i),x,x0(j));
52          vv(j,i) = subs(v(i),x,x0(j));
53          dv(j,i) = diff(v(i),x);
54          dvv(j,i) = subs(dv(j,i),x,x0(j));
55          yy_mul(j,i) = subs(y_mul(i),x0(j));
56          yy_mul2(j,i) = subs(y_mul2(i),x0(j));
57          yy_mul3(j,i) = subs(y_mul3(i),x0(j));
58      end
59  end
60  %%%%%%%%% Plots %%%%%
61  subplot(2,1,1);
62  plot(x0,yy,x0,yy2);
63  subplot(2,1,2);
64  plot(x0,dyy,x0,dyy2);
65  figure;
66  subplot(2,1,1);
67  plot(x0, vv);
68  subplot(2,1,2);
69  plot(x0, dvv);
70  figure;
71  subplot(2,1,1);
72  plot(x0,zz,x0,zz2);
```

```
73  subplot(2,1,2);
74  plot(x0,dzz,x0,dzz2);
75  figure;
76  subplot(2,1,1);
77  plot(x0, uu)
78  subplot(2,1,2);
79  plot(x0, duu)
80  figure;
81  subplot(2,1,1);
82  plot(x0, yy_mul);
83  subplot(2,1,2);
84  plot(x0, yy_mul2,x0, yy_mul3);
```

Table 3.3 Bandwidth consumption (kbps) and download time (s)

Performance	Shaped traffic	Unshaped traffic
HTTP 1 download time	1200	2650
Video streaming 1 buffering	No	Yes
Total bandwidth	758	951

3.3   Resource Allocation: Distributed Architecture and Solution

Section 3.2 introduces a convex utility proportional fairness maximization to perform optimal resource allocation in a single step. The optimization in Sect. 3.2 parameterized subscriber, application status, and QoS as UE subscription weights, application status weights, and application utility functions, respectively. While this centralized architecture assigned application resources in a single step by the BS that served the UEs, hosting the applications, in response to the application utility function parameters transmitted by the host UEs to the BS, the section aims at a distributed architecture for the same resource allocation by assigning application rates in 2 steps from the BS to the UEs and from the UEs to their hosted applications. Additionally, the proposed architecture provides a pricing mechanism to help MNOs flatten their traffic volume in peak-hour times. The centralized architecture for the resource allocation in Sect. 3.2 is based on Eq. (3.10), repeated as Eq. (3.13), below for convenience, in which, as depicted in Fig. 3.1), a BS with R resources assigns rates to M UEs, whose subscription-dependent weight is $\{\beta_i | 1 \leq i leq M\}$, as the rate vector $\mathbf{r} = [r_1, r_2, \ldots, r_M]$.

$$\max_{\mathbf{r}} \quad \prod_{i=1}^{M} \left(\prod_{j=1}^{N_i} U_{ij}^{\alpha_{ij}}(r_{ij}) \right)^{\beta_i}$$

$$\text{subject to} \quad \sum_{i=1}^{M} \sum_{j=1}^{N_i} r_{ij} \le R, r_{ij} \ge 0, \quad i = 1, 2, \ldots, M, \quad j = 1, 2, \ldots, N_i$$

$$(3.13)$$

To cast the formulation in Eq. (3.13) into a distributed architecture, the centralized architecture optimization in Eq. (3.13) is divided into 2 simpler optimization problems to be solved separately. The first optimization, referred to as the External UE Resource Allocation (EURA), gets the BS to assign UE resources via collaborations of the BS and its served UEs, while the second optimization, referred to as the Internal UE Rate Allocation (IURA), has the UEs to internally distribute resources to their hosted applications according to the applications of QoS requirements communicated through their application utility function parameters to the hosting UEs. Moreover, UEs' aggregate utility functions are the same as the centralized architecture, i.e. Eq. (3.14).

$$V_i(r_i) = \prod_{j=1}^{N_i} U_{ij}^{\alpha_{ij}}(r_{ij}) \tag{3.14}$$

EURA optimization can be written as Eq. (3.15) in which $V_i(r_i) = \prod_{j=1}^{N_i} U_{ij}^{\alpha_{ij}}(r_{ij})$ is the ith UE aggregate utility function as expressed in Eq. (3.13), $\mathbf{r} = [r_1, r_2, \ldots, r_M]$ is the UE rate vector with the ith component being the rate assigned by the BS to the ith UE, and M is the number of UEs served by the BS. This section proves that the EURA optimization in Eq. (3.15) is convex, has a tractable optimal solution, and presents an algorithm to solve the EURA optimization in Eq. (3.15).

$$\max_{\mathbf{r}} \quad \prod_{i=1}^{M} V_i^{\beta_i}(r_i)$$

$$\text{subject to} \quad \sum_{i=1}^{M} r_i \le R, r_i \ge 0, \quad i = 1, 2, \ldots, M$$

$$(3.15)$$

IURA optimization is solved internally in each UE and written as Eq. (3.16), for the ith UE with $i \in \{1, 2, \ldots M\}$, in which $\mathbf{r}_i = [r_{i1}, r_{i2}, \ldots, r_{iN_i}]$ is the application rate allocation vector with the jth component being the resources assigned by the ith UE to the jth application, r_i^{opt} is the ith UE resources allocated by the BS by solving the EURA optimization of Eq. (3.15) in the first place, and N_i is the number of applications hosted by the ith UE. Here, the superscript "opt" indicates the optimality of the UE rates. In addition, this section proves that there always exists a tractable global optimal solution to the IURA optimization in Eq. (3.16) and

presents a solution algorithm for the IURA optimization. The next section proves
the convexity of the EURA and IURA optimization problems.

$$\max_{\mathbf{r}_i} \quad \prod_{j=1}^{N_i} U_{ij}^{\alpha_{ij}}(r_{ij})$$

$$\text{subject to} \quad \sum_{j=1}^{N_i} r_{ij} \leq r_i^{\text{opt}}, r_{ij} \geq 0, \qquad j = 1, 2, \ldots, N_i \tag{3.16}$$

Logarithm function is strictly increasing; therefore, an equivalent objective
function $\arg\max_{\mathbf{r}} \sum_{i=1}^{M} \beta_i \log(V_i(r_i))$ to that of the EURA optimization in Eq. (3.15)
is created and referred to as Eq. (3.17) and log-EURA problem.

$$\max_{\mathbf{r}} \quad \sum_{i=1}^{M} \beta_i \log(V_i(r_i))$$

$$\text{subject to} \quad \sum_{i=1}^{M} r_i \leq R, r_i \geq 0, \qquad i = 1, 2, \ldots, M \tag{3.17}$$

Lemma 3.2.1 proves that the aggregate utility natural logarithm $\log(V_i(r_i))$ is
strictly concave. Hence, Theorem 3.3.1 substantiates the convexity of the EURA
optimization in Eq. (3.15), which in turn proves the optimality of the UE rates
assigned by the BS, i.e. the solutions of the EURA optimization.

Theorem 3.3.1 *The EURA optimization in Eq. (3.15) is convex and has a unique
tractable globally optimal solution.*

Proof The aggregate utility concavity from Lemma 3.2.1 reveals the convexity of
the log-EURA optimization [22], which in turn substantiates the convexity of the
EURA optimization in Eq. (3.15) since the objective functions of the log-EURA
and EURA are equivalent. This EURA convexity proves there always exists a unique
tractable globally optimal solution for the EURA optimization [22].

Logarithm function is strictly convex, which means that the IURA objective func-
tion in Eq. (3.16), i.e. $\prod_{j=1}^{N_i} U_{ij}^{\alpha_{ij}}(r_{ij})$, becomes equivalent to $\sum_{j=1}^{N_i} \alpha_{ij} \log(U_{ij}(r_{ij}))$.
Hence, Eq. (3.16) can be expressed as Eq. (3.18), referred to as the log-IURA
optimization, for which Corollary 3.3.2 is provided.

$$\max_{\mathbf{r}_i} \quad \sum_{j=1}^{N_i} \alpha_{ij} \log U_{ij}(r_{ij})$$

$$\text{subject to} \quad \sum_{i=1}^{N_i} r_{ij} \leq r_i^{\text{opt}}, r_{ij} \geq 0, \qquad j = 1, 2, \ldots, N_i \tag{3.18}$$

Corollary 3.3.2 *The IURA optimization in Eq. (3.16) is convex and has a unique tractable globally optimal solution.*

Proof Lemma 3.2.1 proves that natural logarithm of application utility functions provides convex functions that substantiate the convexity of the log-IURA optimization, in Eq. (3.18) [22], which in turn proves the convexity of the IURA optimization in Eq. (3.16) since the log-IURA and IURA optimizations have equivalent objective functions. Therefore, the convex IURA optimization has a tractable globally optimal solution [22].

Since Theorem 3.3.1 and Corollary 3.3.2 indicate the convexity of EURA and IURA optimizations, composing the distributed optimization, the distributed optimization is convex and its solution, i.e. the assigned UE/application rates set, is optimal. The solution to the distributed resource allocation is obtained by applying the Lagrangian of the dual optimizations for the EURA and IURA optimizations. The duality for convex optimization allows them to be solved efficiently [24, 25]. The log-EURA optimization in Eq. (3.17) can be solved by conversion to its dual problem [24, 25]. To do so, the Lagrangian is defined as Eq. (3.19) where $z_i \geq 0$ is the slack variable and p is the Lagrange multiplier of the EURA optimization, referred to as the shadow price (price per unit bandwidth for all the M channels) in the context of the EURA optimization.

$$
L(\mathbf{r}, p) = \sum_{i=1}^{M} \log(V_i(r_i)) - p\left(\sum_{i=1}^{M} r_i + z - R\right) = \sum_{i=1}^{M}\left(\log(V_i(r_i)) - pr_i\right)
$$

$$
+ p(R - z) = \sum_{i=1}^{M} L_i(r_i, p) + p(R - z) \tag{3.19}
$$

Therefore, the ith UE bid for resource assignment can be written as $w_i = pr_i$ in which $\sum_{i=1}^{M} w_i = p \sum_{i=1}^{M} r_i$. The 1st term of Eq. (3.19) is separable in r_i, and one can write $\max_{\mathbf{r}} \sum_{i=1}^{M}(\log(V_i(r_i)) - pr_i) = \sum_{i=1}^{M} \max_{r_i}(\log(V_i(r_i)) - pr_i)$; thus, the dual optimization objective function can be expressed as Eq. (3.20).

$$
D(p) = \max_{\mathbf{r}} L(\mathbf{r}, p) = \sum_{i=1}^{M} \max_{r_i}\left(\log(V_i(r_i)) - pr_i\right) + p(R - z)
$$

$$
= \sum_{i=1}^{M} \max_{r_i}(L_i(r_i, p)) + p(R - z) \tag{3.20}
$$

Next, the dual optimization is formulated as Eq. (3.21).

$$
\min_{p} \quad D(p)
$$
$$
\text{subject to} \quad p \geq 0. \tag{3.21}
$$

Leveraging the Lagrange multipliers, we obtain Eq. (3.22).

$$\frac{\partial D(p)}{\partial p} = R - \sum_{i=1}^{M} r_i - \sum_{i=1}^{M} z_i = 0 \tag{3.22}$$

Substituting $\sum_{i=1}^{M} r_i$ using $\sum_{i=1}^{M} w_i = p \sum_{i=1}^{M} r_i$ yields Eq. (3.23) whose minimum occurs at $p = \frac{\sum_{i=1}^{M} w_i}{R}$, for $z = 0$, in which the ith UE bid sent to the BS for resource assignment is $w_i = p r_i$.

$$p = \frac{\sum_{i=1}^{M} w_i}{R - \sum_{i=1}^{M} z_i} \tag{3.23}$$

Next, we divide the log-EURA optimization in Eq. (3.17) into simpler optimizations at the BS, referred to as the BS EURA optimization, and at the UEs, referred to as the UE EURA optimization as, respectively, expressed in Eqs. (3.25) and (3.24) whose solutions, guaranteeing the utility proportional fairness of Eq. (3.15), are enunciated in Algorithms 4 and 5 correspondingly.

$$\begin{aligned} &\max_{r_i} && \log V_i(r_i) - p r_i \\ &\text{subject to} && p \geq 0, r_i \geq 0, \quad i = 1, 2, \ldots, M \end{aligned} \tag{3.24}$$

The aforesaid algorithms initialize the ith UE bid to $w_i(0) = 0$ and transmits the first bid $w_i(1)$ to the BS that subtracts the received bid $w_i(n)$, where $n = 1, 2, \ldots,$ from the previous bid of the same UE, i.e. $w_i(n-1)$, and either stops if the difference is less than a termination threshold δ or computes and sends the shadow price $p(n) = \frac{\sum_{i=1}^{M} w_i(n)}{R}$ to all its served UEs. Then, the ith UE extracts its rate $r_i(n)$ from the received shadow price $p(n)$ such that $\log V_i(r_i) - p(n) r_i$ is maximized. Next, the aforesaid rate $r_i(n)$ is used to estimate the new bid $w_i(n) = p(n) r_i(n)$, transmitted to the BS. This routine repeats until the UE bid difference $|w_i(n) - w_i(n-1)|$ falls below the threshold δ. The solution of the ith UE EURA optimization, i.e. $r_i(n) = \arg\max_{r_i} \left(\log V_i(r_i) - p(n) r_i \right)$, in Algorithm 4 essentially solves the equation $\frac{\partial \log V_i(r_i)}{\partial r_i} = p(n)$, which is the Lagrange multiplier for Eq. (3.24), whose geometrical interpretation is the intersection of the horizontal line $y = p(n)$ with the curve $y = \frac{\partial \log V_i(r_i)}{\partial r_i}$.

$$\begin{aligned} &\min_{p} && D(p) \\ &\text{subject to} && p \geq 0. \end{aligned} \tag{3.25}$$

Algorithm 4 UE EURA optimization algorithm

1: Send initial bid $w_i(1)$ to BS.
2: **loop**
3:     Receive shadow price $p(n)$ from BS.
4:     **if** STOP from eNB **then**
5:        Calculate allocated rate $r_i^{\text{opt}} = \frac{w_i(n)}{p(n)}$.
6:        STOP
7:     **else**
8:        Solve $r_i(n) = \arg\max_{r_i}\Big(\log V_i(r_i) - p(n)r_i \Big)$.
9:        Send new bid $w_i(n) = p(n)r_i(n)$ to BS.
10:    **end if**
11: **end loop**

Algorithm 5 BS EURA optimization algorithm

1: **loop**
2:     Receive bids $w_i(n)$ from UEs. {Let $w_i(0) = 1$ $\forall i$}
3:     **if** $|w_i(n) - w_i(n-1)| < \delta$ $\forall i$ **then**
4:        Allocate rates, $r_i^{\text{opt}} = \frac{w_i(n)}{p(n)}$ to user i.
5:        STOP
6:     **else**
7:        Calculate $p(n) = \frac{\sum_{i=1}^{M} w_i(n)}{R}$.
8:        Send new shadow price $p(n)$ to all UEs.
9:     **end if**
10: **end loop**

To start a convergence analysis of the EURA Algorithms 4 and 5, Lemma 3.2.2, which says that the aggregate utility function $V_i(r_i)$ and the slope curvature function $\frac{\partial \log V_i(r_i)}{\partial r_i}$ have inflection points at $r_i = r_{ij}^s \approx r_{ij}^{\text{inf}}$ for jth application utility function U_{ij} and are convex for $r_{ij} > \max_j r_{ij}^s$, is helpful. Next, Corollary 3.3.3 is considered.

Corollary 3.3.3 *If* $\sum_{i=1}^{M} \max_j r_{ij}^s \ll R$, *then Algorithms 4 and 5 converge to globally optimal rates corresponding to the steady state shadow price* $p_{ss} < \frac{a_{i_{\max}} d_{i_{\max}}}{1 - d_{i_{\max}}} + \frac{a_{i_{\max}}}{2}$, *where* $i_{\max} = \arg\max_i r_{ij_{\max}}^s$ *and* $r_{ij_{\max}}^s = \max_j r_{ij}^s$.

Proof To reach the optimal solution of Algorithm 4, one should solve $r_i(n) = \arg\max_{r_i}\Big(\log V_i(r_i) - p(n)r_i \Big)$ via the Lagrange multipliers in Eq. (3.26).

$$\frac{\partial \log V_i(r_i)}{\partial r_i} - p = S_i(r_i) - p = 0. \tag{3.26}$$

Furthermore, Eq. (3.7) shows that the slope curvature function $S_i(r_i)$ is convex for $r_i > \max_j r^s_{ij} \approx \max_j b_{ij}$. Thus, Algorithms 4 and 5 are guaranteed to converge to the globally optimal solution when $S_i(r_i)$ is in its convex region. Hence, the aggregate utility function natural logarithm also converges to the globally optimal solution for $r_i > \max_j r^s_{ij} \approx \max_j b_{ij}$. On the other hand, the sigmoidal utility function's inflection point occurs at $r^{\inf}_i = b_{ij}$. For $\sum_{i=1}^{M} \max_j r^s_{ij} \ll R$, Algorithms 4 and 5 allocate rates $r_{ij} > b_{ij}$ for all users, and since $S_{ij}(r_{ij})$ is convex for $r_{ij} > r^s_{ij} \approx b_{ij}$, the optimal solution can be obtained by the algorithms. Equation (3.26) and convexity of $S_{ij}(r_{ij})$ for $r_{ij} > r^s_{ij} \approx b_{ij}$ reveal that $p_{ss} < S_{ij}(r_{ij} = \max b_{ij})$, where $S_{ij}(r_{ij} = \max b_{ij}) = \frac{a_{i\max} d_{i\max}}{1-d_{i\max}} + \frac{a_{i\max}}{2}$ and $i_{\max} = \arg\max_i b_{ij}$.

Next, Corollary 3.3.4 illustrates that the EURA optimization can be unstable.

Corollary 3.3.4 *For $\sum_{i=1}^{M} \max_j r^s_{ij} > R$ and the globally optimal shadow price*

$$p_{ss} \approx \frac{a_{ij} d_{ij} e^{\frac{a_{ij} b_{ij}}{2}}}{1-d_{ij}(1+e^{\frac{a_{ij} b_{ij}}{2}})} + \frac{a_{ij} e^{\frac{a_{ij} b_{ij}}{2}}}{(1+e^{\frac{a_{ij} b_{ij}}{2}})}, \text{ the solution by EURA Algorithms 4 and 5}$$

fluctuates about the globally optimal solution.

Proof It follows from Lemma 3.2.2 that for $\sum_{i=1}^{M} r^{\inf}_{ij} > R$, $\exists i$ such that the optimal rates $r^{\text{opt}}_{ij} < b_{ij}$. Thus, $p_{ss} \approx \dfrac{a_{ij} d_{ij} e^{\frac{a_{ij} b_{ij}}{2}}}{1-d_{ij}(1+e^{\frac{a_{ij} b_{ij}}{2}})} + \dfrac{a_{ij} e^{\frac{a_{ij} b_{ij}}{2}}}{(1+e^{\frac{a_{ij} b_{ij}}{2}})}$ is the optimal shadow price for the optimization in Eq. (3.15). Then, a small change in the shadow price $p(n)$ at the nth iteration causes the rate $r_{ij}(n)$—the root of $S_{ij}(r_{ij}) - p(n) = 0$—to fluctuate between the concave and convex curvatures of the slope curve $S_{ij}(r_{ij})$ for the ith UE. Therefore, it produces a fluctuation in the bid value $w_i(n)$ sent to the BS that creates a vacillation of the shadow price $p(n)$ sent from the BS to the UEs. Therefore, the iterative solution oscillates about the global optimal rates r^{opt}_{ij}.

The EURA optimization instability in Corollary 3.3.4 means non-optimality of the rates assigned by the EURA optimization as illustrated in Theorem 3.3.5.

Theorem 3.3.5 *EURA Algorithms 4 and 5 do not converge to the optimal solution for all BS rates R.*

Proof Corollaries 3.3.3 and 3.3.4 reveal that the EURA algorithm does not converge to the globally optimal solution for all values of R.

Adding robustness to the EURA Algorithms 4 and 5 so it converges for all BS maximum resource values requires the algorithm not to fluctuate for $\sum_{i=1}^{M} \max_j r^s_{ij} \ll R$. To do this, a *fluctuation decay function* $\Delta w(n)$, which is either an exponential fluctuation decay function $\Delta w(n) = l_1 e^{-\frac{n}{l_2}}$ or a rational fluctuation decay function $\Delta w(n) = \frac{l_3}{n}$ where l_1, l_2, l_3 can be adjusted to change the bid w_i decay rate, reduces the step size between two consecutive bids of an ith user, i.e. $w_i(n) - w_i(n-1)$, for whom fluctuations occur. The allocated rates should have synergy with those of EURA Algorithms 4 and 5 for $\sum_{i=1}^{M} \max_j r_{ij}^s > R$. The fluctuation decay function can be included in either UE EURA Algorithm 4 or BS EURA Algorithm 5. In our model, we incorporate the decay function into the UE EURA Algorithm 4 even though it can be incorporated into the BS EURA Algorithm 5 too. The fledgling robust EURA Algorithms 6 and 7 start with $w_i(0) = 0$, after which each UE transmits an initial bid $w_i(1)$ to its serving BS that, at the nth iteration, calculates the difference between the two latest consecutive bids, i.e. $w_i(n) - w_i(n-1)$, and either stops when the difference is smaller than a termination threshold δ or computes the shadow price $p_E(n) = \frac{\sum_{i=1}^{M} w_i(n)}{R}$ and sends it to its served UEs, and the ith UE extracts its rate r_i, maximizes the statement $\log \beta_i V_i(r_i) - p_E(n)r_i$, estimates a new bid $w_i(n) = p_E(n)r_i(n)$, and sends it to the BS. This procedure is repeated until two consecutive bids' difference for the ith UE be smaller than a termination threshold δ.

Algorithm 6 UE EURA robust algorithm

1: Send initial bid $w_i(1)$ to BS.
2: **loop**
3:     Receive shadow price $p(n)$ from BS.
4:     **if** STOP from eNB **then**
5:         Calculate rate $r_{ij}^{\text{opt}} = \frac{w_i(n)}{p(n)}$.
6:         STOP
7:     **else**
8:         Solve $r_i(n) = \arg\max_{r_i}\left(\beta_i \log V_i(r_i) - p_E(n)r_i \right)$.
9:         Calculate new bid $w_i(n) = p(n)r_i(n)$.
10:        **if** $|w_i(n) - w_i(n-1)| > \Delta w(n)$ **then**
11:            $w_i(n) = w_i(n-1) + \text{sign}(w_i(n) - w_i(n-1))\Delta w(n)$
12:        **end if**
13:        Send new bid $w_i(n)$ to BS.
14:    **end if**
15: **end loop**

Algorithm 7 BS EURA algorithm

1: **loop**
2:     Receive bids $w_i(n)$ from UEs. {Let $w_i(0) = 1 \ \forall i$}
3:     **if** $|w_i(n) - w_i(n-1)| < \delta \ \forall i$ **then**
4:         Allocate rates, $r_i^{\text{opt}} = \frac{w_i(n)}{p(n)}$ to user i.
5:         STOP and allocate rates (i.e r_i^{opt} to user i)
6:     **else**
7:         Calculate $p_E(n) = \frac{\sum_{i=1}^{M} w_i(n)}{R}$
8:         Send new shadow price $p_E(n)$ to all UEs
9:     **end if**
10: **end loop**

Remark 3.3.6 If the subscriber differentiation parameter β_i is available only at the BS, the shadow price p_E is changed to $\frac{p_E}{\beta_i}$.

Next, we explain the solution to the second stage of the distributed resource allocation where the application rates r_{ij} are assigned optimally by the UEs via Algorithm 8, where the ith UE leverages the EURA allocated optimal rate r_i^{opt} to solve $\mathbf{r}_i = \arg\max_{\mathbf{r}_i} \sum_{j=1}^{N_i} (\alpha_{ij} \log U_{ij}(r_{ij}) - p_I r_{ij}) + p_I r_{ij}^{\text{opt}}$.

Algorithm 8 UE IURA algorithm

---

loop

  Receive r_i^{opt} from eNB. {by EURA Algorithms}

  Solve

  $\mathbf{r}_i = \arg\max_{\mathbf{r}_i} \sum_{j=1}^{N_i} (\alpha_{ij} \log U_{ij}(r_{ij}) - p_I r_{ij}) + p_I r_i^{\text{opt}}$ {Note: $\mathbf{r}_i = \{r_{i1}, r_{i2}, \ldots, r_{iN_i}\}$}

  Allocate r_{ij} to the jth application.

end loop

3.3.1 Distributed Resource Allocation Simulation

Similar to the simulations of the centralized architecture in Sect. 3.2.1, a cell with $M = 6$ UEs and a BS as in Fig. 3.1 is considered and the UEs simultaneously run a delay-tolerant and a real-time application modeled, respectively, as logarithmic and sigmoidal application utility functions whose parameters are in Table 2.1. The sigmoidal application utility function with parameters $a = 5$, $b = 10$ represents a step function at rate $r = 5$, while parameters $a = 3$, $b = 15$ represent a real-time application utility function with an inflection point at rate $r = 15$, whereas application utility function parameters $a = 1$, $b = 25$ depict real-time applications with the inflection point $r = 25$, and the logarithmic utility functions with $r^{\max} = 100$ and distinct k_i parameters represent delay-tolerant FTP applications. The plots of the application utility functions in Table 2.1 are shown in Fig. 2.1 from which we can observe that the real-time applications require a minimum rate equal to the inflection point after which the application QoS is mostly fulfilled. On the other hand, the logarithmic utility function is provided with marginal QoS at even low rates. Furthermore, Fig. 2.1 shows, as expected, that the utility functions are strictly increasing and zero valued at zero rates. Moreover, the first derivative of the utility function natural logarithm, $S_{ij}(r_{ij})$, is depicted in Fig. 3.3, which reflects that the first derivative is positive and decreasing in compliance with Lemma 3.2.3. Then, the distributed resource allocation solution Algorithms 6, 7, and 8 were applied to the logarithmic and sigmoidal utility functions whose usage percentage is the application status weight vector in Eq. (3.14) as $\alpha = \{\alpha_{11}, \alpha_{21}, \alpha_{31}, \alpha_{41}, \alpha_{51}, \alpha_{61}, \alpha_{12}, \alpha_{22}, \alpha_{32}, \alpha_{42}, \alpha_{52}, \alpha_{62}\}$. Here, α_{ij} is the status weight of the jth application of the ith UE.

In addition, the aggregate utility functions $V_i(r_i)$ for $i \in \{1, \dots, 6\}$ are illustrated in Fig. 3.7a. In the following simulations, the termination threshold is $\delta = 10^{-4}$ and the BS resources R sweep from 10 to 200 in a step size of 5 bandwidth units. Also, the application status weight vector is $\alpha = \{0.1, 0.5, 0.9, 0.1, 0.5, 0.9, 0.9, 0.5, 0.1, 0.9, 0.5, 0.1\}$. For the distributed resource allocation Algorithms 6, 7, and 8, UE resources are allocated according to Fig. 3.4a in the EURA optimization algorithm with the changes in R. Initially, all UEs are allocated rates as they have real-time applications requiring an immediate resource allocation before any QoS is met. As a case in point, UE2 has a real-time streaming video application (Table 2.1) requiring resource assignment right away. The rates are equal to those of the centralized resource allocation; this observation is in line with the conclusion in Sect. 3.4 according to which both centralized and distributed approaches produce identical rates. Figure 3.7b shows the UE bids $\{w_i | i \in \{1, \dots, 6\}\}$ in the EURA optimization algorithm under changing BS R. First, more resources become available at the BS, which are equivalent to higher UE resources being assigned. Nonetheless, a small R causes the UEs with applications with stringent QoS requirements to bid higher so as to gain resources. For example, since UE2 subsumes a real-time streaming video application, it initially bids higher and receives a fast resource allocation as in Fig. 3.4a. Then, the UEs employ the distributed resource allocation IURA optimization to internally assign resources to their applications according to the application bids. Figure 3.4b depicts the allocated application rates $\{r_{ij} | i \in \{1, \dots, 6\} \wedge j \in \{1, 2\}\}$ obtained from the IURA optimization as a function of the variations in the BS resources R and reveals that initially more resources are given to the real-time applications due to their stringent QoS requirements. The application bids $\{w_{ij} | i \in \{1, \dots, 6\} \wedge j \in \{1, 2\}\}$ in the IURA optimization as a function of the changes in the BS maximum resources R. Since the real-time applications need more resources, they bid higher than the delay-tolerant ones when the resources are scarce specially. Moreover, such applications with higher QoS requirements as the real-time streaming video of UE1, the red plot, bid even higher to gain more resources. It is noteworthy that the abundance of BS resources R reduces the bid values.

Since the utility proportional fairness objective functions are employed in the formation of the distributed resource allocation optimizations in Eqs. (3.15) and (3.16), the algorithms do not assign a zero rate to any UE; thereby, a minimum QoS is guaranteed. Besides, a BS assigns a majority of its resources to the real-time applications until they receive their utility inflection rate $r_{ij} = b_{ij}$. However, when the total BS resources exceed the addition of the inflection point rates $\sum b_{ij}$, the BS can give more resources to the delay-tolerant applications as well; this behavior is observed with the assigned rate increase and bid value decrease that occur after the BS resources R exceed $R = \sum b_{ij} = 105$ [10]. In addition, the improvement in the robust Algorithms 6 and 7 is revealed in the shadow price fluctuation reduction shown in Fig. 3.8 where the decay function stabilizes the resource allocation by doing away with oscillations. This behavior is similarly observed for Algorithms 6 and 7 over Algorithms 4 and 5 when $R > \sum b_{ij} = 105$, yet Algorithms 6 and 7 fail to allocate optimal rates for $R < \sum b_{ij} = 105$. Hence, Algorithms 6 and 7

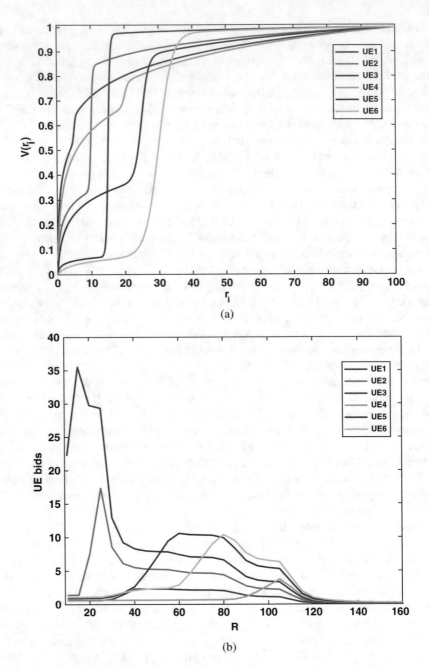

(a)

(b)

Fig. 3.7 (a) shows the aggregate utility functions and multiplication of application utility functions $V_i(r_i)$ powered to the usage percentages vs. the UE rates r_i for $i \in \{1, \ldots, 6\}$. (b) depicts the UE bids to obtain resources

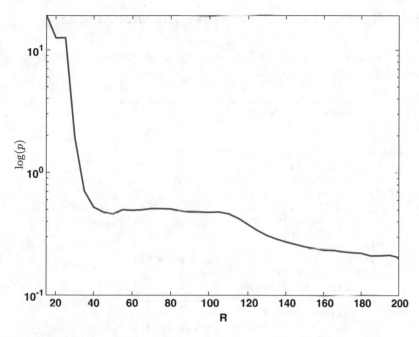

Fig. 3.8 Shadow price p vs. BS resources R shows that availability of more BS resources reduces the shadow price

are robust even under a scarce resource availability. For example, the real-time applications modeled by the sigmoidal application utility functions bid higher/lower when the BS resources are scarce/abundant; thus, a pricing proportional to the bids that outfits MNOs with a pricing mechanism to increase the service price for under high load traffic circumstances to motivate mobile subscribers to use the network during off-peak hours. The shadow price p(n), price per unit bandwidth for all users and applications vs. the BS maximum resources is depicted in Fig. 3.8 which depicts a high price for small R in high traffic situations with a constant number of users and small price for large R values in low-traffic circumstances with the same number of users. The large plummets in the shadow price after the rates $R = \{15, 25, 85, 105\}$, the inflection point of the sigmoidal utility functions, are notable. Additionally, a large shadow price decline is visible at $\sum_{i=1}^{k} r_{ij}^{\inf}$, where $k = \{1, 2, \ldots, M\}$ is the UE index for M UEs and i is the user with the maximum utility function slope $\arg\max_i S_i(r_i)$, in our case UE 3 ($b_{3j} = 15$), followed by UE 2 ($b_{2j} = 10$), then, UE 1, UE 5, and UE 6 that have $S_i(r_i)$ ($b_{1j} = 5, b_{5j} = 25, b_{6j} = 30$), and ultimately UE 4 ($b_{4j} = 20$). The larger the slope difference $\Delta S_{ij} = |S_i(r_i) - S_j(r_j)|$, the more the variation of the shadow price p(n) vs. R.

3.3.2   MATLAB Code for Distributed Resource Allocation

This is the MATLAB code used to generate the simulations of the distributed resource allocation in Sect. 3.3.1.

```
1   %1. Main Code
2   syms x
3   global p_old
4   NoUE = 6;     % number of users
5   for i_rate = 1:1:39
6       Rate(i_rate) = 5 + i_rate * 5 % Maximum Rate
7       w = [10 10 10 10 10 10 ];
8       r_min = [0 0 0 0 0 0 ];          % initial new min_rate ...
            of optimization
9       Δ = 1;
10      time = 0;
11      %%%%%%%% UE Rates   %%%%%%%%%%%%%%%%%%%%
12      while  (Δ > 0.01)
13          time = time + 1;
14          w_old = w;
15          t(time) = time;
16          p(time) = eNodeB(w, Rate(i_rate));   % sent from ...
                eNodeB
17          for i = 1: NoUE
18              pp = p(time);
19              ii =i;
20              soln(i) = fzero(@(x) ...
                    utility_UE(x,ii,pp),[.001 1000]);
21              r_opt(i) = max(soln(i), r_min(i));
22              w(i) = r_opt(i) * p(time);
23              if abs(w_old(i)-w(i)) > (5.* ...
                    exp(-0.1*time))%(10 ./ time)
24                  w(i) = w_old(i) + (5.* exp(-0.1*time)) .* ...
                        sign(w(i)-w_old(i));
25              end
26          end
27          w_sim(time,:) = w;
28          r_sim(time,:) = w_sim(time,:)./p(time);
29          Δ = max(abs(w-w_old));
30      end
31      t_final =time;
32      r_ss(i_rate,:) = r_sim(time,:);
33      w_ss(i_rate,:) = w_sim(time,:);
34      p_ss(i_rate) = p(time);
35      time_ss(i_rate) = time;
36      time_ss2(i_rate) = 1;
37      %%%%%%%% Application Rates   %%%%%%%%%%%%%%%%%%%%
38      wUE = [10 10 10 10 10 10 10 10 10 10 10 10];
39      w_oldUE = [100 100 100 100 100 100 100 100 100 100 ...
            100 100];
40      r_minUE = [0 0 0 0 0 0 0 0 0 0 0 0];
```

```
41        ΔUE = 1;
42        for i = 1: NoUE
43            time = 0;    % initate for each UE
44            while  (time<40)
45                time = time + 1;
46                t(time) = time;
47                ii =i;
48                ii2 =i + NoUE;
49                w_oldUE(ii) = wUE(ii);
50                w_oldUE(ii2) = wUE(ii2);
51                R2 = r_sim(t_final, ii);
52                pUE(time,ii) = UE(ii, wUE, R2);
53                pUE(time,ii2) = pUE(time,ii);
54                ppUE(ii) = pUE(time,ii);
55                soln(ii) = fzero(@(x) ...
                      utility_UE(x,ii,ppUE(ii)),[.0001 100]);
56                r_optUE(ii) = max(soln(ii), r_minUE(ii));
57                wUE(ii) = r_optUE(ii) * ppUE(ii);
58                soln(ii2) = fzero(@(x) ...
                      utility_UE(x,ii2,ppUE(ii)),[.0001 100]);
59                r_optUE(ii2) = max(soln(ii2), r_minUE(ii2));
60                wUE(ii2) = r_optUE(ii2) * ppUE(ii);
61                if abs(w_oldUE(ii)-wUE(ii)) > (5.* ...
                      exp(-0.1*time))
62                    wUE(ii) = w_oldUE(ii) + (5.* ...
                          exp(-0.1*time)) .* ...
                          sign(wUE(ii)-w_oldUE(ii));
63                end
64                if abs(w_oldUE(ii2)-wUE(ii2)) > (5.* ...
                      exp(-0.1*time))
65                    wUE(ii2) = w_oldUE(ii2) + (5.* ...
                          exp(-0.1*time)) .* ...
                          sign(wUE(ii2)-w_oldUE(ii2));
66                end
67                w_simUE(time,ii) = wUE(ii);
68                w_simUE(time,ii2) = wUE(ii2);
69                r_simUE(time,ii) = ...
                      w_simUE(time,ii)./pUE(time,ii);
70                r_simUE(time,ii2) = ...
                      w_simUE(time,ii2)./pUE(time,ii2);
71                ΔUE = max(abs(wUE-w_oldUE));
72            end
73        end
74        r_ssUE(i_rate,:) = r_simUE(time,:);
75        w_ssUE(i_rate,:) = w_simUE(time,:);
76    end
77    %%%%%%%%% Plot UE Rate%%%%%%%%%%%%%%%%%
78    plot(Rate,r_ss)
79    xlabel('Rate');
80    legend('UE1','UE2','UE3','UE4','UE5','UE6');
81    ylabel('UE rates')
82    %%%%%%%%% Plot UE Bids%%%%%%%%%%%%%%%%%
83    figure;
84    plot(Rate,w_ss)
```

```
85  xlabel('Rate');
86  legend('UE1','UE2','UE3','UE4','UE5','UE6');
87  ylabel('UE bids ')
88  figure;
89  plot(Rate,time_ss,Rate,time_ss2)
90  xlabel('Rate');
91  ylabel('Iterations')
92  %%%%%%%% Plot Application Rate%%%%%%%%%%%%%%%%
93  figure;
94  plot(Rate,r_ssUE)
95  xlabel('Rate');
96  legend('UE1 App1','UE1 App1','UE2 App1','UE2 App1','UE3 ...
        App1','UE3 App1');
97  ylabel('App rates')
98  %%%%%%%% Plot Application Rate%%%%%%%%%%%%%%%%
99  figure;
100 plot(Rate,w_ssUE)
101 xlabel('Rate');
102 legend('UE1 App1','UE1 App1','UE2 App1','UE2 App1','UE3 ...
        App1','UE3 App1');
103 ylabel('App bids')
104 figure;
105 plot(Rate,time_ssUE);
106 %2. Called by Main
107 function [p] = eNodeB(w)
108 global p_old R
109 R = 180;
110 p = sum(w)/R;
111 %3. Called by Main
112 function [p] = eNodeB(w,Rate)
113 R = Rate;
114 p = sum(w)/R;
```

3.4   Mathematical Equivalence

Here, we show the mathematical equivalence of the distributed resource allocation
with the centralized one. Figure 3.9 shows that the BS in the centralized architecture
receives the QoS requirements from the UEs for each of their hosted applications,
and then the BS assigns the rates to all the applications of the UEs accordingly. On
the other hand, the problem can be simplified by decomposing the application rate
allocation into 2 simpler optimizations in which first the BS assigns rates to the UEs
that in turn allocate resources to the applications. Next, Corollary 3.4.1 proves that
the centralized resource allocation in Eq. (3.13) yields the same UE and application
rates as the distributed resource allocation in Eqs. (3.15) and (3.16) in order to show
their equivalence.

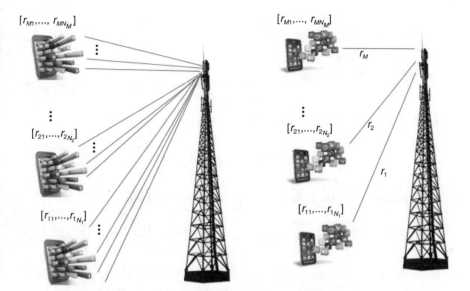

Fig. 3.9 The centralized resource allocation assigns application resources after receiving the request from UE for all applications. This architecture can be decomposed to BS and UE optimizations where the BS assigns UE resources on a holistic basis and the UE allocates application rates

Corollary 3.4.1 *The optimal rates assigned by the distributed resource allocation in Eq. (3.15) and (3.16) are equal to the one allocated by the centralized resource allocation in Eq. (3.13).*

Proof The centralized resource allocation optimization in Eq. (3.13) can be written as the log-centralized optimization in Eq. (3.12) that is Eq. (3.27). The Lagrangian of the log-centralized optimization can be expressed as Eq. (3.28), where $z \geq 0$ is the slack variable and p_T is the Lagrange multiplier.

$$\max_{\mathbf{r}} \quad \sum_{i=1}^{M} \beta_i \sum_{j=1}^{N_i} \alpha_{ij} \log U_{ij}(r_{ij})$$

$$\text{subject to} \quad \sum_{i=1}^{M} \sum_{j=1}^{N_i} r_{ij} \leq R, r_{ij} \geq 0, \quad i = 1, 2, \ldots, M, \, j = 1, 2, \ldots, N_i$$

$$(3.27)$$

To solve the Lagrangian, we take its derivative with respect to $\mathbf{r}$ as shown in Eq. (3.28).

$$L_T(\mathbf{r}) = \left(\sum_{i=1}^{M} \beta_i \sum_{j=1}^{N_i} \alpha_{ij} \log U_{ij}(r_{ij}) \right) - p_T \left(\sum_{i=1}^{M} \sum_{j=1}^{N_i} r_{ij} - R + z \right)$$

$$\Rightarrow \frac{\partial L_T(\mathbf{r})}{\partial r_{ij}} = 0 \Rightarrow p_T = \beta_i \alpha_{ij} S_{ij}(r_{ij}) \tag{3.28}$$

So, the ith UE's jth application rate can be written as Eq. (3.29).

$$r_{ij} = S_{ij}^{-1}\left(\frac{p_T}{\beta_i \alpha_{ij}}\right) \tag{3.29}$$

Using Eq. (3.3), the ith UE rate can be written as Eq. (3.30).

$$N_i p_T = \beta_i S_i(r_i) \Rightarrow r_i = S_i^{-1}\left(\frac{N_i p_T}{\beta_i}\right) \tag{3.30}$$

Next, let us focus on the distributed resource allocation. The Lagrangian of the distributed resource allocation log-EURA optimization can be written as Eq. (3.31), where $z \geq 0$ is the slack variable and p_E is the Lagrange multiplier. To solve the Lagrangian, we take its derivative with respect to $\mathbf{r}$.

$$L_E(\mathbf{r}) = \left(\sum_{i=1}^{M} \beta_i \log V_i(r_i) \right) - p_E \left(\sum_{i=1}^{M} r_i - R + z \right) \Rightarrow \frac{\partial L_E(\mathbf{r})}{\partial r_i}$$

$$= \beta_i S_i(r_i) - p_E = 0 \Rightarrow p_E = \beta_i S_i(r_i) \tag{3.31}$$

So, the ith UE rate can be written as

$$r_i = S_i^{-1}\left(\frac{p_E}{\beta_i}\right) \tag{3.32}$$

Replacing S_i from Eq. (3.3), we can write Eq. (3.33).

$$p_E = \beta_i \sum_{j=1}^{N_i} \alpha_{ij} S_{ij}(r_{ij}) \tag{3.33}$$

And, we get Eq. (3.34) below.

$$p_E = \sum_{j=1}^{N_i} p_T = N_i p_T \tag{3.34}$$

Equation (3.30) gives the UE rate r_i for the centralized resource allocation equal to that of the distributed resource allocation EURA optimization (which gives the

UE rates) as in Eq. (3.32); this signifies that the centralized and distributed resource allocations produce identical UE rates. Next, we focus on the application rates. The Lagrangian of the distributed resource allocation log-IURA optimization in Eq. (3.18) can be expressed as Eq. (3.35), where $z \geq 0$ is the slack variable and p_I is the Lagrange multiplier, essentially the UE shadow price (price per bandwidth unit) for all applications in the ith UE.

$$L_I(r_i) = \left(\sum_{j=1}^{N_i} \alpha_{ij} \log U_{ij}(r_{ij}) \right) - p_I \left(\sum_{j=1}^{N_i} r_{ij} - r_{ij}^{opt} + z \right) \Rightarrow \frac{\partial L_I(r_i)}{\partial r_{ij}}$$

$$= \alpha_{ij} S_{ij}(r_{ij}) - p_I = 0 \Rightarrow p_I = \alpha_{ij} S_{ij}(r_{ij}) \qquad (3.35)$$

Adding the ith UE applications gives Eq. (3.36).

$$\sum_{j=1}^{N_i} p_I = \sum_{j=1}^{N_i} \alpha_{ij} S_{ij}(r_{ij}) \qquad (3.36)$$

Using Eq. (3.3), we can then write Eq. (3.37).

$$\beta_i N_i p_I = \beta_i S_i(r_i) = p_E = N_i p_T \Rightarrow p_T = \beta_i p_I \qquad (3.37)$$

Then, the ith UE assigns a rate in Eq. (3.38) to its jth application.

$$r_{ij} = S_{ij}^{-1} \left(\frac{p_I}{\alpha_{ij}} \right) = S_{ij}^{-1} \left(\frac{p_T}{\beta_i \alpha_{ij}} \right) \qquad (3.38)$$

Considering the constraints of the Eq. (3.18), the total rate of the ith UE can be written as Eq. (3.39).

$$r_i^{opt} = \sum_{j=1}^{N_i} r_{ij} = \sum_{j=1}^{N_i} S_{ij}^{-1} \left(\frac{p_T}{\beta_i \alpha_{ij}} \right) \qquad (3.39)$$

Equation (3.38) gives an application rate r_{ij} from the distributed resource allocation IURA optimization equal to that of the centralized resource allocation in Eq. (3.29); this signifies that the centralized and distributed resource allocations produce equal application rates. Thus, the distributed resource allocations in Eqs. (3.15) and (3.16) and the centralized resource allocation in Eq. (3.13) yield identical UE and application rates.

Next, Theorem 3.4.2 proves that the centralized resource allocation in Eq. (3.13) is mathematically equivalent to the distributed resource allocation in Eqs. (3.15) and (3.16).

Theorem 3.4.2 *The distributed resource allocation in Eqs. (3.15) and (3.16) is equivalent to the centralized resource allocation in Eq. (3.13).*

Proof Corollary 3.4.1 proves that the distributed and centralized resource allocation methods produce identical UE and application resources. Therefore, the distributed resource allocation in Eqs. (3.15) and (3.16) is equivalent to the centralized resource allocation in Eq. (3.13).

3.5 Distributed or Centralized

This section provides a brief discussion on whether the distributed resource allocation or the centralized resource allocation is preferable. The answer depends on the situation to which the resource allocation gets applied. For the current-day wireless systems with smartphones with capabilities and battery life as of this book publication, the centralized resource allocation is preferable. The distributed resource allocation requires the implementation of the IURA in the UEs. Performing these calculations at resource allocation periods of a few milliseconds, common in LTE and 5G scheduling systems, are too taxing on smartphones and consumes the battery life of the devices. This situation is further exacerbated by implementing the robust version of the distributed resource allocation under a rebidding regime. As we mentioned earlier, the assigned resources remain optimal under a rebidding policy, and having the UEs to consistently compute resources, send bids, and so forth further deteriorates the battery of the smartphones. On the other hand, 5G relies heavily on software implementation, *aka* virtualization, which shifts the burden of computations to the cloud whose enormous computation capability significantly helps the resource allocation implementation. Such a cloud-based architecture renders itself conveniently to the centralized implementation of the resource allocation for cellular communication systems. In view of this discussion, the centralized resource allocation is more conducive to the cellular communication systems' resource allocation as it stands today. On the other hand, as we move toward 5G, new services such as Multi-access Edge Computing (MEC) [7, 12, 17, 33–35] can decentralize the computations like the resource allocation. In such a scenario, the calculations to assign resources to the UEs can be shifted to the MEC unit. Such a framework is obviously conducive to a distributed architecture. However, since the resources are requested by UEs, while MEC and 5G can operate via the distributed resource allocation, the question boils down to the allocation between the MEC and UEs for which a centralized approach is still preferential in order to refrain from having the UEs to perform calculations and, thereby, adversely impacting their battery life and users' QoE.

3.6 Benchmark Comparison

In order to evaluate our proposed resource allocation modus operandi with other proportional fairness methods, we select the approach in [38] that uses the same optimization as in Eq. (3.40) with weighted logarithm functions, provided in

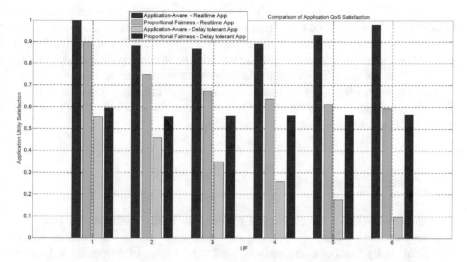

Fig. 3.10 Comparison of the resources allocated by the method in Sect. 3.3.1 and that of [38] in Eq. (3.41)

Eq. (3.41), instead of sigmoidal utility functions. The weight factors w_i essentially endeavor to fit the logarithmic utility functions to sigmoidal ones in order to use convexity in solving the optimization.

$$\max_{\mathbf{r}} \quad \prod_{i=1}^{M} \left(\prod_{j=1}^{N_i} U_{ij}^{\alpha_{ij}} (r_{ij}) \right)^{\beta_i}$$

$$\text{subject to} \quad \sum_{i=1}^{M} \sum_{j=1}^{N_i} r_{ij} \leq R, r_{ij} \geq 0, \quad i = 1, 2, \ldots, M, \quad j = 1, 2, \ldots, N_i$$

$$(3.40)$$

To find the weights of the logarithm functions, we use Levenberg–Marquardt algorithm (LMA) [39, 40] whose primary application is in the least square curve fitting. Essentially, for n datum pairs of independent/dependent variables (x_i, y_i), the LMA optimizes parameters β of the function $f(x; \beta)$ so that it minimizes $\sum_{i=1}^{n} (y_i - f(x_i; \beta))$.

$$U_i(r_i) = \chi_i log(r_i) \tag{3.41}$$

To do the comparison, we use the logarithmic and sigmoidal utility functions as in Sect. 3.3.1 but curve-fit the sigmoidal utility functions by the LMA to the weighted logarithm function in Eq. (3.41). The results are depicted in Fig. 3.10 which illustrates the resources allocated to the real-time applications by the method in Sect. 3.3.1 and by the method in [38] as respectively the dark blue and light blue bars. Moreover, the resources given to the delay-tolerant applications by the method in Sect. 3.3.1 and by the method in [38] are shown respectively with yellow and brown bars. The figure shows that the method in [38] gives real-time applications at least 10% less resources than that of the method in Sect. 3.3.1, whereas the former gives more resources to the delay-tolerant traffic in spite of the real-time applications needing more resources compared to the delay-tolerant traffic.

This shows that the proposed method in Sect. 3.3.1 is application-aware since it allocates resources based on the traffic QoS requirements because it gives more/less resources to real-time/delay-tolerant applications.

3.7 Chapter Summary

This chapter leveraged the concept of application utility functions for modeling QoS requirement of delay-tolerant and real-time applications via logarithmic and sigmoidal utility functions, respectively, and the concept of proportional fairness optimization to create a QoS-minded utility proportional fairness resource allocation that incorporated usage and subscriber differentiation and was cast under centralized architecture. This chapter proved the convexity of the proposed centralized resource allocation, solved it through the Lagrangian of the dual problem, and depicted simulations to reveal the effect of the application of the proposed centralized resource allocation. Finally, this chapter implemented the centralized architecture in a real-world network to show that the absence of the resource allocation caused video streaming buffering and so it reduced QoE, while the presence of the resource allocation eliminated any video streaming buffering by lengthening the delay-tolerant applications' download time and so it increased the QoE without hurting applications' QoS requirements. Furthermore, this chapter developed distributed resource allocation architecture composed of the EURA optimization that allocated the UE rates by the BS and the IURA optimization that assigned application rates by the UEs. Not only did this chapter prove the convexity of the proposed distributed resource allocation, but it also proved that the distributed resource allocation does not lead to optimal rates for all resource availability situations at the BS and introduced a variation of the distributed resource allocation that incorporated robustness into the distributed resource allocation via decay functions. Finally, this chapter performed simulations to show the application of the proposed distributed resource allocation to a cellular communication system.

References

1. Ericsson mobility report, in *Ericsson Press Release*, 2013
2. H. Ekstrom, QoS control in the 3GPP evolved packet system. IEEE Commun. Mag. **47**(2), 76–83 (2009)
3. A. Kumar, A. Abdelhadi, T.C. Clancy, A delay efficient multiclass packet scheduler for heterogeneous M2M uplink, in *IEEE MILCOM*, 2016
4. Machina Research, M2M growth necessitates a new approach to network planning and optimisation (2015). https://goo.gl/CDmd6M
5. A. Sengupta, A. Abdelhadi, T.C. Clancy, Performance trade-offs in IoT uplink networks under secrecy constraints (under submission)
6. A. Kumar, A. Abdelhadi, T.C. Clancy, An online delay efficient packet scheduler for M2M traffic in industrial automation, in *IEEE Systems Conference*, 2016
7. M. Ghorbanzadeh, Y. Chen, K. Ma, C. Clancy, R. McGwier, A neural network approach to category validation of Android applications, in *IEEE Conference on Computing, Networking, and Communications (ICNC)*, 2013
8. Z. Kbah, A. Abdelhadi, Resource allocation in cellular systems for applications with random parameters (2015). abs/1507.07608
9. Q. Fang, *Distinctions Between Levenberg–Marquardt Method and Tikhonov Regularization* (Dartmouth College Publication, 2004).
10. M. Ghorbanzadeh, A. Abdelhadi, C. Clacy, *Cellular Communications Systems in Congested Environments Resource Allocation and End-to-End Quality of Service Solutions with MATLAB* (Springer, Berlin, 2017)
11. M. Ghorbanzadeh, Resource allocation and end-to-end quality of service for cellular communications systems in congested and contested environments. Ph.D. Thesis, Virginia Tech, 2015
12. M. Ghorbanzadeh, Y. Chen, C. Clancy, Fine-grained end-to-end network model via vector quantization and hidden Markov processes, in *IEEE Conference on Communications (ICC)*, 2013
13. M. Ibrahim, Channel quality indicator feedback in long term evolution (LTE) system. IOSR J. Electron. Commun. Eng. **9**(2), 14—19 (2014)
14. G. T. S. . R. 8, Multiplexing and channel coding (FDD)
15. 3rd Generation Partnership Project, Evolved universal terrestrial radio access (e-ULTRA): physical layer procedures (release 8), in *TS* 2008
16. X. Li, Q. Fang, L. Shi, A effective SINR link to system mapping method for CQI feedback in TD-LTE system, in *2011 IEEE 2nd International Conference on Computing, Control and Industrial Engineering*, vol. 2 (2011), pp. 208–211
17. M. Ghorbanzadeh, A. Abdelhadi, C. Clancy, A utility proportional fairness radio resource block allocation in cellular networks, in *IEEE International Conference on Computing, Networking and Communications (ICNC)*, 2015
18. M. Ghorbanzadeh, A. Abdelhadi, C. Clancy, A utility proportional fairness bandwidth allocation in radar-coexistent cellular networks, in *Military Communications Conference (MILCOM)*, 2014
19. A.J. Goldsmith, S.-G. Chua, Adaptive coded modulation for fading channels. IEEE Trans. Commun. **46**(5), 595–602 (1998)
20. D. Kim, B.C. Jung, H. Lee, D. K. Sung, H. Yoon, Optimal modulation and coding scheme selection in cellular networks with hybrid-ARQ error control. IEEE Trans. Wirel. Commun. **7**(12), 5195–5201 (2008)
21. J. Fan, Q. Yin, G.Y. Li, B. Peng, X. Zhu, MCS selection for throughput improvement in downlink LTE systems, in *2011 Proceedings of 20th International Conference on Computer Communications and Networks (ICCCN)* (IEEE, Piscataway, 2011), pp. 1–5
22. S. Boyd, L. Vandenberghe, *Introduction to Convex Optimization with Engineering Applications* (Cambridge University Press, Cambridge, 2004)

23. Y. Chen, M. Ghorbanzadeh, K. Ma, C. Clancy, R. McGwier, A hidden Markov model detection of malicious Android applications at runtime, in *2014 23rd Wireless and Optical Communication Conference (WOCC)*, 2014
24. F. Kelly, A. Maulloo, D. Tan, Rate control in communication networks: shadow prices, proportional fairness and stability. J. Oper. Res. Soc. **49**(3), 237–252 (1998)
25. S. Low, D. Lapsley, Optimization flow control, I: basic algorithm and convergence. IEEE/ACM Trans. Netw. **7**(6), 861–874 (1999)
26. M. Portnoy, *Virtualization Essentials* (Wiley, London, 2012)
27. IBM, x3250 m4. IBM Redbooks Product Guide, 2011
28. A. Bachmutsky, *System Design for Telecommunication Gateways* (Wiley, 2010)
29. C. Spurgeon, *Ethernet: The Definitive Guide* (O'Reilly, 2014)
30. W. Stallings, Data and computer communications, in *William Stallings Books on Computer and Data Communications*, 2013
31. M. Ghorbanzadeh, E. Visotsky, P. Moorut, W. Yang, C. Clancy, Radar inband and out-of-band interference into LTE macro and small cell uplinks in the 3.5 GHz band, in *2015 IEEE Wireless Communications and Networking Conference (WCNC)*, 2015
32. M. Ghorbanzadeh, E. Visotsky, P. Moorut, W. Yang, C. Clancy, Radar in-band interference effects on macrocell LTE uplink deployments in the U.S. 3.5 GHz band, in *2015 International Conference on Computing, Networking and Communications (ICNC)*, 2015
33. M. Ghorbanzadeh, E. Visotsky, P. Moorut, W. Yang, C. Clancy, Radar interference into LTE base stations in the 3.5 GHz band. Phys. Commun. **20**, 33–47 (2016)
34. H. Shajaiah, M. Ghorbanzadeh, A. Abdelhadi, C. Clancy, Application-aware resource allocation based on channel information for cellular networks, in *2019 IEEE Wireless Communications and Networking Conference (WCNC)* (2019), pp. 1–6
35. M. Ghorbanzadeh, A. Abdelhadi, C. Clancy, Application-aware resource allocation of hybrid traffic in cellular networks. IEEE Trans. Cogn. Commun. Netw. **3**(2), 226–241 (2017)
36. L. Chappell, G. Combs, *Wireshark Network Analysis: The Official Wireshark Certified Network Analyst Study Guide* (Protocol Analysis Institute, Chappell University, 2010)
37. A. Abdelhadi, T. Clancy, An optimal resource allocation with frequency reuse in cellular networks (2015). abs/1507.07161
38. I.-H. Hou, C.S. Chen, Self-organized resource allocation in LTE systems with weighted proportional fairness, in *2012 IEEE International Conference on Communications (ICC)*, 2012
39. K. Levenberg, A method for the solution of certain non-linear problems in least squares. Q. Appl. Math. **2**(2), 164–168 (1944)
40. D. Marquardt, An algorithm for least-squares estimation of nonlinear parameters. SIAM J. Appl. Math. **11**(2), 431–441 (1963)

Chapter 4
Radio Resource Block Allocation

4.1 Introduction

Resource allocation methods aiming at fulfilling the QoS requirements of modern networks have been the focus of many research studies [1–8]. This attention to QoS-minded resource allocation techniques is partly because modern cellular systems have smart phones running applications with distinct QoS requirements simultaneously. While many of the state-of-the-art resource allocation studies [9–12] consider the traffic nature, these do not address all t QoS-related issues simultaneously. Chapter 3 developed a single-carrier radio resource allocation formulation under a utility proportional fairness framework, proved the optimality of the resource allocation method, incorporated the traffic nature, subscriber type, and application usage percentages in the formulation, and cast the resource allocation under a centralized and distributed architecture. This chapter complements the centralized and distributed optimal resource allocation developed in Chap. 3 by diving into radio resource block (RRB), a discretization of the resource allocation formulations presented in Chap. 3. Section 4.2 discusses the relaxation of the continuous optimization referenced in Chap. 3 and provides the RRB allocation formulation and simulation thereof, and Sect. 4.4 summarizes the chapter.

© The Author(s), under exclusive license to Springer Nature Switzerland AG 2022
M. Ghorbanzadeh, A. Abdelhadi, *Practical Channel-Aware Resource Allocation*,
https://doi.org/10.1007/978-3-030-73632-3_4

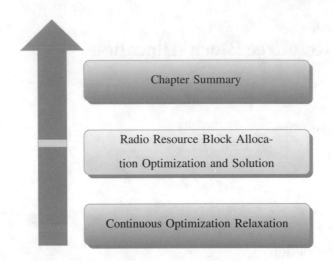

4.2 Continuous Optimization Relaxation

This section expands the proportional fairness resource allocation in Chap. 3 to an RRB allocation for cellular communications with the UEs running delay tolerant and real-time applications whose QoS is modeled by sigmoidal and logarithmic application utility functions, respectively. The optimization as shown in Chap. 3 maximizes the proportional fairness of the system utility functions, while it distributes RRBs to the UEs efficiently. The RRB allocation is achieved by means of the Lagrangian relaxation [5, 6, 13–16] of a convex continuous resource allocation optimization (Chap. 3) by formulating the RRB assignment as an integer nonlinear resource allocation optimization whose positive integer solutions are the RRBs to be distributed by a network scheduler. The Lagrangian relaxation first solves the continuous resource allocation optimization by leveraging its dual optimization, Chap. 3. Then, the discrete boundary points of the continuous optimal resources reflect the possible candidate solutions, i.e. RRBs, to be assigned by the BS to its served UEs. After qualified candidate solutions are found, those that maximize utility function proportional fairness are chosen as the selected solutions. This approach warrants a minimum QoS as the proportional fairness formulation never allocates zero resources. A cell with a single BS and M UEs as in Fig. 4.1 is considered. The resource allocated by the BS to the ith UE is referred to as r_i and the UE utility function is denoted as $U_i(r_i)$, logarithmic and sigmoidal functions with Eq. (4.1), respectively, modeling the QoS satisfaction percentage as a function of r_i. The objective is finding the RRBs which the BS allocates under a proportional fairness policy in order to guarantee a minimum QoS for the UEs. In Eq. (4.1), $I_A(x)$ is an indicator function which is 1 when $x \in A$ and 0 otherwise. Thus, the first term of the addition has $I_{\text{Sigmoidal}} = 1$ when the utility function is a sigmoidal and 0 otherwise, while the second term of the addition has $I_{\text{Logarithmic}}$ when the utility function is logarithmic and 0 otherwise.

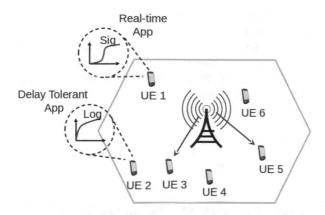

Fig. 4.1 System Model: Single cell, within the cellular network, with a BS covering $M = 6$ UEs, each with simultaneously running delay-tolerant and relay-time applications represented by logarithmic and sigmoidal utility functions, respectively

$$U_i(r_i|x) = c_i\left(\frac{1}{1 + e^{-a_i(r_i-b_i)}} - d_i\right)I_{\text{Sigmoidal}}(x) + \frac{\log(1 + k_i r_i)}{\log(1 + k_i r_{max})}I_{\text{Logarithmic}}(x)$$

(4.1)

Moreover, $c_i = \frac{1+e^{a_i b_i}}{e^{a_i b_i}}$ and $d_i = \frac{1}{1+e^{a_i b_i}}$, and $U(0) = 0$ and $\lim_{r_i \to \infty} U(r_i|$ Sigmoidal$) = 1$ between which the latter indicates a 100% QoS satisfaction for infinite resources. Chapter 2 discussed that b_i is the inflection point, for the sigmoidal utility function of Eq. (4.1), beyond which the QoS satisfied and before which the QoS is adversely affected. The various parameters a_i, b_i, c_i, and d_i shape directly the utility function and are equivalent to various real-time applications. In Eq. (4.1), the parameter r_{max} is the rate to achieve 100% QoS satisfaction, and k_i is the rate with which the utility percentage increases with enlarging r_i. Also, it can be verified $U(r_{max}|$Logarithmic$) = 1$ meaning that, after being provided with a certain amount of resources, the application receives an acceptable QoS satisfaction. Chapter 3 put the proportional fairness resource allocation as Eq. (4.2), where $\mathbf{r} = [r_1, r_2, \ldots, r_M]$ is a rate vector specifying the allocation to the M UEs by their serving BS. The resource allocation objective function assigns BS resources to the UEs such that the resource allocation maximizes the objective function while maintaining the proportional fairness of the individual utility functions. Chapter 3 mentioned that such a resource allocation formulation causes nonzero resource assignment to all the UEs and, as a result, guarantees a minimum QoS. Moreover, Chap. 3 explained that this method assigns more resources to UEs with real-time applications; however, in order to tailor the resource allocation method to wireless technologies such as LTE and 5G, which distribute discrete resources and not continuous ones as radio resource elements (RREs), RB assignments in Eq. (4.2) are proposed. Since the BS has limited resources R, we modify the formulation in Eq. (4.2) to account for these 2 factors. The bandwidth available at the BS is a constraint of the optimization in Eq. (4.2), while RB allocations can be assigned by making another nonlinear constraint that the resources (solutions) are positive integer numbers. Adding these constraints to the optimization in Eq. (4.2) creates a discrete resource allocation optimization. The next section presents the formulation of the discrete RB allocation optimization.

$$\max_{\mathbf{r}} \prod_{i=1}^{M} U_i(r_i) \tag{4.2}$$

4.3 Radio Resource Block Allocation Optimization and Solution

The utility proportional RB allocation optimization is shown in Eq. (4.3), where R is the total resources available to the BS serving the M UEs and $\mathbf{r} = [r_1, r_2, \ldots, r_M]$ is the UE RB vector such that r_i is the RB allocated to the ith UE whose QoS is represented by the application utility function $U_i(r_i|x)$ with $x =$ Sigmoidal, Logarithmic for a sigmoidal and logarithmic application utility function. It is conspicuous that the nonlinear constraint in Eq. (4.3) is a modification of the optimizations presented in Chap. 3.

$$\max_{\mathbf{r}} \quad \prod_{i=1}^{M} U_i(r_i)$$
$$\text{subject to} \quad \sum_{i=1}^{M} r_i \leq R, r_i \geq 0, r_i \in \mathbb{N}, \quad i = 1, 2, \ldots, M \tag{4.3}$$

Chapter 3 proved the concavity of the logarithms of the sigmoidal and logarithmic utility functions in Eq. (4.1), the convexity of the continuous optimization in Eq. (4.3), and the existence/tractability of a globally optimal solution for the optimization whose rates are not positive integers. Relaxing the discreteness constraint from our RB allocation optimization in Eq. (4.3) renders the solution easily by generating the continuous rates, global solutions for which the proportional utility fairness is maximized. Next, Sect. 4.2 explains a modus operandi to change the discrete optimization problem in Eq. (4.3) to another optimization with continuous rates by eliminating the $r_i \in \mathbb{N}$ constraint. Relaxing the $r_i \in \mathbb{N}$ constraint of the RB allocation in Eq. (4.3) via the Lagrangian relaxation [13, 17], Eq. (4.4) is the utility proportional fairness continuous rate allocation optimization. Temporary deletion of the constraint $r_i \in \mathbb{N}$ of Eq. (4.3) yields in another easy-to-solve optimization.

$$\max_{\mathbf{r}} \quad \prod_{i=1}^{M} U_i(r_i)$$

$$\text{subject to} \quad \sum_{i=1}^{M} r_i \leq R, r_i \geq 0, \quad i = 1, 2, \ldots, M \tag{4.4}$$

To obtain the Lagrangian multiplier of the dual optimization of Eq. (4.4), we first define the Lagrangian as Eq. (4.5) in which $z \geq 0$ is the slack variable of the Lagrangian and p is the Lagrange multiplier. Chapter 3 depicted that p is the total price per unit bandwidth for the M UEs.

$$L(\mathbf{r}, p) = \sum_{i=1}^{M} \log(U_i(r_i)) - p\left(\sum_{i=1}^{M} r_i + z - R\right)$$

$$= \sum_{i=1}^{M} \left(\log(U_i(r_i)) - pr_i\right) + p(R - z) \tag{4.5}$$

Then, $w_i = pr_i$ is the ith UE's bid, and we can write it as $\sum_{i=1}^{M} w_i = p \sum_{i=1}^{M} r_i$. Next, the dual optimization is written in Eq. (4.6).

$$\min_{p} \max_{\mathbf{r}} L(\mathbf{r}, p)$$

$$\text{subject to} \quad p \geq 0 \tag{4.6}$$

It is noticeable that the separability of $\sum_{i=1}^{M}(\log(U_i(r_i)) - pr_i)$ in r_i stems out $\max_{\mathbf{r}} \sum_{i=1}^{M}(\log(U_i(r_i)) - pr_i) = \sum_{i=1}^{M} \max_{r_i}(\log(U_i(r_i)) - pr_i)$, and the dual problem in Eq. (4.6) can be expressed as follows:

$$\min_{p} \sum_{i=1}^{M} \max_{r_i}\left(\log(U_i(r_i)) - pr_i\right) + p(R - z) \tag{4.7}$$

$$\text{subject to} \quad p \geq 0$$

We can solve the optimization in Eq. (4.7) by Lagrange multipliers. Since $\frac{\partial \max_{\mathbf{r}} L}{\partial p}$ and $\sum_{i=1}^{M} w_i = p \sum_{i=1}^{M} r_i$, Eq. (4.8) is resulted.

$$p = \frac{\sum_{i=1}^{M} w_i}{R - z} \tag{4.8}$$

It is notable that the shadow price p is minimized when $z = 0$ so that $\sum_{i=1}^{M} w_i = pR$. In essence, Eq. (4.4) is divided into 2 simpler optimizations for the UEs and the BS in a similar fashion to Chap. 3. Then, the ith UE optimization can be expressed as

$$\max_{r_i} \quad \log U_i(r_i) - pr_i$$

$$\text{subject to} \quad p \geq 0, r_i \geq 0, \quad i = 1, 2, \ldots, M \tag{4.9}$$

Besides, the BS optimization can be expressed as Eq. (4.10). The solution $r_i(n) = \arg\max_{r_i}\left(\log U_i(r_i) - p(n)r_i\right)$ is the value of r_i that solves the equation $\frac{\partial \log U_i(r_i)}{\partial r_i} = p(n)$, geometrically the intersection of the horizontal line $y = p(n)$ with the curve $y = \frac{\partial \log U_i(r_i)}{\partial r_i}$ for the ith UE. The utility proportional fairness in the objective function of the optimization in Eq. (4.4) is guaranteed for the UE optimization in Eq. (4.9) and also for BS optimization in Eq. (4.10). The solution algorithms of these optimizations are given in Algorithms 9 and 10 whose development was done in Chap. 3 for the continuous resource allocation.

$$\min_{p} \max_{\mathbf{r}} L(\mathbf{r}, p)$$

$$\text{subject to} \quad p \geq 0 \tag{4.10}$$

Algorithms 9 and 10 start with each UE sending an initial bid $w_i(1)$ to its serving BS which calculates the difference between the previous two received bids. Should the difference absolute value be less than a threshold δ, the algorithms terminate their operation and exit; otherwise the BS calculates a new shadow price by equation $p(n) = \frac{\sum_{i=1}^{M} w_i(n)}{R}$ and sends it to its served UEs. Then, the UEs use the shadow price to find the continuous rate r_i, which maximizes $\log U_i(r_i) - p(n)r_i$, employed to calculate a new bid $w_i(n) = p(n)r_i(n)$ sent to the BS. Like Chap. 3, the robust algorithm to ensure that convergence of the continuous rate allocation algorithm for all BS bandwidths R is used. A convergence measure $w(n) == l1e^{-\frac{n}{l_2}}$ is used to reduce the step size between consecutive bids. Here, adjusting the parameters l_1, l_2, l_3 or $\Delta w(n) = \frac{l_3}{n}$ makes the bids and rates change accordingly. The first decay function yields in an exponential decrease. The fluctuation decay function can be included in either Algorithm 9—in the UE—or Algorithm 10—in the BS optimization. We have placed the decay function in the UE Algorithm 9 in this section. The shadow price $p(n)$ fluctuation decreases in each iteration n, and it converges to an optimal shadow price germane to the optimal rates.

Algorithm 9 UE algorithm (from Chap. 3)

1: Send initial bid $w_i(1)$ to BS.
2: **loop**
3:      Receive shadow price $p(n)$ from BS.
4:      **if** STOP from BS **then**
5:          Calculate the rate $r_i^{\text{opt}} = \frac{w_i(n)}{p(n)}$.
6:          STOP
7:      **else**
8:          Calculate new bid $w_i(n) = p(n)r_i(n)$.
9:          **if** $|w_i(n) - w_i(n-1)| > \Delta w(n)$ **then**
10:             $w_i(n) = w_i(n-1) + \text{sign}(w_i(n) - w_i(n-1))\Delta w(n)$.
11:         **end if**
12:         Send new bid $w_i(n)$ to BS.
13:     **end if**
14: **end loop**

Algorithm 10 BS algorithm (from Chap. 3)

1: **loop**
2:      Receive bids $w_i(n)$ from UEs. $\{$Let $w_i(0) = 1$ $\forall i\}$
3:      **if** $|w_i(n) - w_i(n-1)| < \delta$ $\forall i$ **then**
4:          STOP and calculate rates $r_i^{\text{opt}} = \frac{w_i(n)}{p(n)}$.
5:      **else**
6:          Calculate $p(n) = \frac{\sum_{i=1}^{M} w_i(n)}{R}$.
7:          Send new shadow price $p(n)$ to all UEs.
8:      **end if**
9: **end loop**

After the continuous rates from Algorithms 9 and 10 are found for the equivalent relaxed continuous rate allocation optimization in Eq. (4.4), we focus on finding discrete resources as solutions to the RB allocation optimization in Eq. (4.3). The next section presents an approach to map the continuous resources to feasible discrete RBs. This solves the discrete optimization in Eq. (4.3) based upon a sequence of feasible candidates extracted from the optimal continuous resources. According to Chap. 3, the application utility functions are logarithmic for the delay-tolerant traffic and sigmoidal for the real-time traffic, and the logarithm of both of which is concave; this means that the optimization has a globally optimal solution, i.e. rates assigned to UEs by the BS. For example, posit that the proportional fairness optimization for an LTE system with 1 BS and M UEs is $U(\mathbf{r}) = U(r_1, \ldots, r_M) = \prod_{i=1}^{M} U_i(r_i)$, which is proven to have a global optimal solution $[r_{c1}, \ldots, r_{cM}]$ where $r_{c1}, \ldots, r_{cM}] = \arg\max_{\mathbf{r}} \prod_{i=1}^{M} U_i(r_i)$. $[r_c 1, \ldots, r_c M]$ is the continuous optimal rate vector obtained from Algorithms 9 and 10. Next, we consider all the discrete points in the M dimensional domain of $U(r_1, \ldots, r_M)$ as the Cartesian product of the possible values. For example, should the values of UE resources r_i range from 1 to 100, there exist 100^M feasible discrete rates. Obviously, an exhaustive search of the possible discrete resources is computationally complex as the size of the system grows. To surmount this complexity, we can focus on the neighbors of the continuous optimal rates, i.e. for each component of the rate vector, the lower (floor function) integer points less than the continuous rate and the upper (ceiling function) integer points higher than the continuous rate are chosen. This reduces the number of possibilities for the discrete rate candidates to 2^M sequences, of much less computation complexity as opposed to the original 100^M points. Such an approach corresponds to locating the floors and ceilings of the continuous rates, i.e. $[\lfloor r_{c1} \rfloor, \ldots, \lfloor r_{cM} \rfloor]$ and $[\lceil r_{c1} \rceil, \ldots, \lceil r_{cM} \rceil]$. The candidate discrete rates are the M dimensional vectors whose elements are the floors or ceilings of the equivalent continuous rates, which yields in 2^M feasible RB vectors. For a system of only 2 UEs, the utility function is a surface of 2 dimensional points with real number coordinates r_1 and r_2, equivalent to the continuous rates given to $U_1(r_1)$ and $U_2(r_2)$. The rates must be transformed to discrete points, so the RB allocation is pragmatic. Even though there are many valid discrete points on the r_i axes in the $U(r_1, r_2)$ domain, concentrating on the boundary points for $[r_1, r_2]$ gives 4 integer points as $[\lfloor r_1, \lfloor r_2 \rfloor], [\lceil r_1, \lfloor r_2 \rfloor], [\lfloor r_1, \lceil r_2 \rceil],$ and $[\lceil r_1 \rceil r_2]$. The boundary points (the closest integers to the continuous rates r_1 and r_2) are 4 combinations (4 discrete rates in red, green, yellow, and cyan), the closest to the continuous point in blue. Consequently, the search space confines to only the 4 boundary points as opposed to all possible discrete points on the surface of the wireless system utility function. To further restrict the search space, we consider the continuous optimization constraint $\sum_{i=1}^{M} r_i \leq R$. While the aforementioned boundary points are the closest to the optimal continuous solution, they might infringe this constraint due to floor and ceiling operations. Thus, we evaluate and select the candidate boundary points satisfying this requirement.

The methodology that extracts the RB candidates from the continuous ones obtained from the optimization in Eq. (4.4) is shown in Algorithm 11. As we

---

Algorithm 11 Resource block allocation

---

loop
  Continuous value of r_i is r_i^{opt}. {Map the floors of continuous rates less than unity to one unit.}
  Calculate the ceilings and floors of the r_i.
end loop
List all possible sequences that can be obtained from the floors and ceilings.
if sum of the discrete rates surpasses R **then**
  Eliminate that sequence of discrete rates.
else
  Store that sequence of discrete rates as a candidate for RB allocation.
end if
loop
  Calculate the system utility for the stored RB candidates.
  Store RBs which maximize the utility.
end loop

---

observe, after locating the RB candidates, we evaluate the optimization utility for each candidate and store those that maximize the utility function. Algorithm 11 maps the continuous rates that are smaller than one unit bandwidths to 1 both in the ceiling and in the floor as an RB allocation of 1 unit is the minimum possible discrete resource allocation under a proportional fairness policy which avoids from zero resource allocation. It is interesting to mention that many applications do not require many resources and for these the discrete rates may be mapped to the RB 1. Such an approach stems out equal discrete sequences for a continuous rate sequence. On the other hand, this algorithm can generate numerous RB sequences, which provides the BS with more options and a higher flexibility in assigning the resources. This is particularly of high consequence in scenarios with several BSs as it can decrease the intra-cell and inter-cell interferences by having BSs giving distinct RBs to the UEs. The next section portrays the simulation results for Algorithms 10, 9, and 11.

4.3.1   Radio Resource Block Allocation Simulation Results

This section investigates Algorithms 10, 9, and 11 applied to a representative scenario with 6 UEs and 1 BS in Fig. 4.1. Each UE runs 1 application such that 3 of the applications are delay-tolerant and the other 3 are real-time applications. The sigmoidal utility functions with parameters $a = 5$, $b = 10$ to model VoIP, $a = 3$, $b = 20$ to demonstrate a real-time standard streaming video, and $a = 1$, $b = 30$ to depict an HDV application and logarithmic utility functions with $k = \{15, 3, 0.5\}$ to model FTP services are considered. The utility functions vs. RBs are shown in Fig. 4.2a. The 6 plotted application utility functions correspond to the applications running on the 6 UEs as illustrated in Fig. 4.1 where 6 UEs are served by the BS. UEs 1, 2, and 3 are running real-time applications represented by sigmoidal utility functions, and UEs 4, 5, and 6 have delay-tolerant applications modeled by logarithmic utilities. The goal is for the BS to assign RBs to the UEs. The red sigmoidal function in Fig. 4.2a shows the HDV application (inflection point 30). The green sigmoidal function models the standard streaming video, and the blue one represents a VoIP application. Figure 4.2a shows that, after surpassing the inflection points, utility QoS percentages are about 100% (ref. Chap. 2). The purple, cyan, and yellow logarithmic utility functions are FTP applications, respectively, sorted in the ascending order of their delay-tolerance behavior. For example, for an identical rate, the utility function shown by the yellow curve is under the other logarithmic function curves, which depicts a lower QoS satisfaction. We relaxed the discrete RB allocation optimization to an easy-to-solve continuous rate allocation optimization. The utility functions equivalent to the RB allocation in the relaxed continuous resource allocation are illustrated in Fig. 4.3b. Comparing Figs. 4.2a and 4.3b, we see that the continuous optimization relaxation generates utility functions following the discrete ones closely such that the RB allocation resembles samples from the continuous version of the resource allocation optimization.

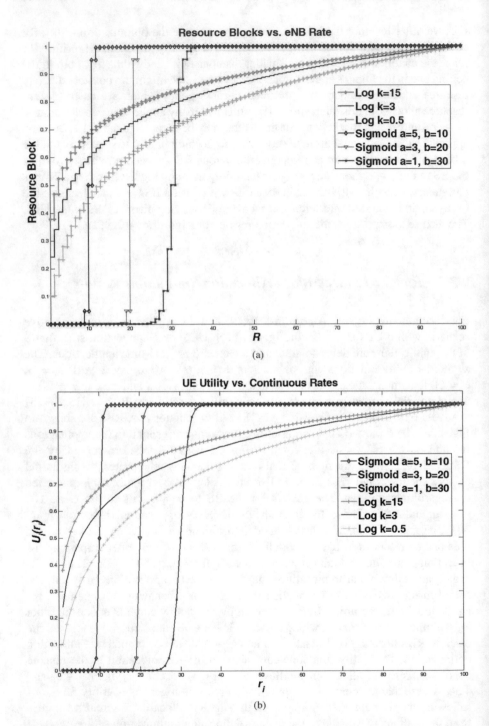

Fig. 4.2 (a) Represents utility function of one of the UEs; 3 of whose applications are delay-tolerant and 3 others are real-time applications. (b) Shows the continuous optimization utility functions where the horizontal axis is the continuous rates. (a) RRBs vs total resources available. (b) Continuous optimization utility functions

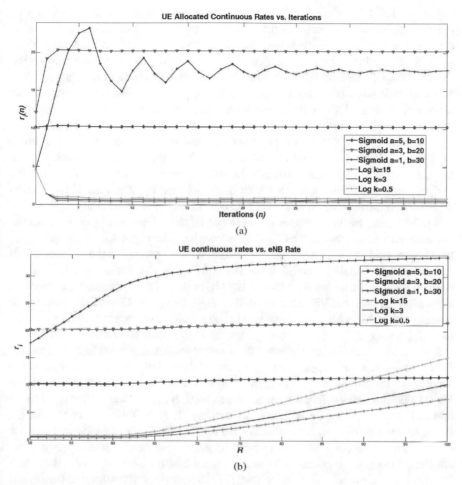

Fig. 4.3 (**a**) Shows relaxed optimization continuous resource allocation: red, green, and blue curves represent VoIP, standard streaming video, and HDV applications on UEs 1, 2, and 3. Also, cyan, purple, and yellow curves model the resources assigned for FTP applications on UEs 4, 5, and 6. (**b**) Shows relaxed optimization continuous resource allocations vs. the BS bandwidth

Here, we consider two scenarios, one where for $R = 50$ and another one where $50 \leq R \leq 100$. In the first scenario, the BS resource availability is $R = 50$ and $n = 35$. Figure 4.3a shows the continuous resources assigned to the 6 UEs vs. the iterations of the algorithm for the relaxed continuous resource allocation optimization and that the sigmoidal application utility functions are prioritized by the algorithm over the logarithmic utility functions. It is noticeable that after the resource allocation is stable, the resources are above inflection points of the sigmoidal utility functions; however, this depends on whether there are enough resources available at the BS to meet the demand portrayed by the inflection points of the sigmoidal utility functions; otherwise, the applications QoS are gravely impacted. The UE bids/resources are proportional during the various iterations of the algorithms, which means that the larger bids produce higher UE rates. Likewise, the delay-tolerant applications bid higher at the commence of the algorithms initially. To evaluate the performance of the proposed method in allocating RBs, the following simulation with 40 iterations and the BS rate R in the range of 50–100 with unit steps is considered. Similar to Fig. 4.3a, BS distributes most of the resources to those UEs running real-time applications. The procedure continues until these applications' utility functions reach their inflection points; yet, if R is less than the sum of the inflection points, the real-time UEs obtain less resources than their inflection rate, and this adversely affects the application QoS. On the flip side, when the BS rate R exceeds the sum of the inflection point rates, the BS assigns more resources to the UEs with delay-tolerant applications. This situation occurs when the bandwidth of the BS exceeds 65 units in this simulation.

Furthermore, we observe that the UEs bid as the BS resources change. As the BS resources grow, the UEs bid smaller as depicted in Fig. 4.4a, where the larger bids correspond to higher rates and UE bids decrease severely as the BS bandwidth increases. Additionally, increasing the BS bandwidth R reduces the UE bids and shadow price as the higher R, the lower the UE bids, and the lower the shadow price. Figure 4.4b shows that RBs are allocated to the UEs as the BS available bandwidth changes using Algorithm 11. Since most modern communication networks such as the LTE leverage discrete rates, i.e. RBs, to assign the radio resources, it is the RB allocations that renders the continuous rate resource allocation to be pragmatic. UEs running real-time applications get more RBs initially when the resources are scarce. On the flip side, increasing R enables the BS to assign more RBs to the UEs which run delay-tolerant applications as well. Next, we look at the continuous and RB allocations for an arbitrary BS bandwidth R in Table 4.1 of which, for example, the first column is an arbitrary BS bandwidth R, the second column is a 6 dimensional continuous rate vector whose ith element is the resource assigned to the ith UE by the relaxed continuous rate allocation optimization, and the last column is a discrete 6 dimensional vector whose ith element is the RB allocated the ith UE using Algorithm 11. As we can see, an optimal continuous rate can be mapped to one or more RBs, which gives the BS the flexibility to select RBs from a pool of feasible discrete allocations. This is helpful when more than 1 BS is in the system to ameliorate inter-cell and intra-cell interference .

As we can see in Table 4.1, for a BS bandwidth of 100, the relaxed continuous rate allocation optimization assigns resources to 6 UEs optimally as [11.57, 21.57, 33.58, 7.72, 10.36, 15.21], maximizing the utility proportional fairness. On the other hand, for this BS rate, 8 possible RB allocations are feasible; these are [11, 21, 33, 8, 11, 16], [11, 21, 33, 8, 11, 15], [11, 21, 33, 8, 10, 16], [11, 21, 33, 8, 10, 15], [11, 21, 33, 7, 11, 16], [11, 21, 33, 7, 11, 16], [11, 21, 33, 7, 10, 16], and [11, 21, 33, 7, 10, 15], and each of them is a possible RB allocation; however, a BS rate of 50 yields in a continuous resource allocation [10.46, 20.46, 24.21, 9.13, 9.13, 9.13] and discrete RB allocation [10, 20, 17, 1, 1, 1]. Ultimately, let us consider the computational complexity of the RB Algorithm 11. Since each continuous rate r_i assigned to the ith UE's

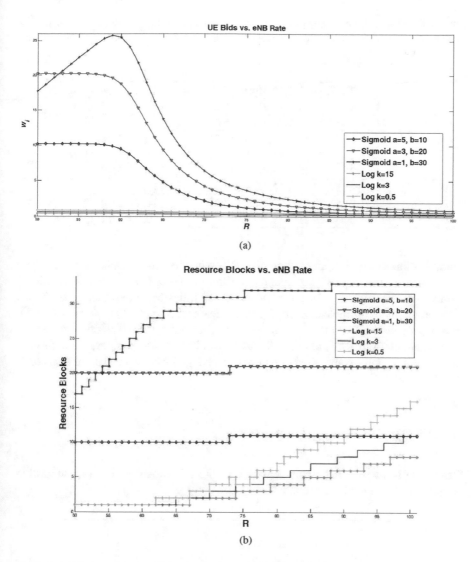

Fig. 4.4 (**a**) Shows relaxed optimization UE bids vs. the BS bandwidth from which we see as more resources become available to the BS, applications bid lower to have bandwidths allocated to them. (**b**) Shows the discrete optimization RRB allocation vs. the BS bandwidth, where the red, blue, and green curves are bids for UEs, which run VoIP, HDV, and standard streaming video applications, and cyan, yellow, and purple curves are FTP applications bids

utility function, U_i has at minimum 1 and at maximum 2 possible boundary points. Hence, the computational complexity of the algorithm is at most $O(2^M)$ for M UEs. This is not all 2^M possible combinations of the lower bounds and upper bounds of the continuous rates; thus, some of the initially feasible discrete RB candidates are not included as they violate the BS available resources in the

Table 4.1 Continuous Rate Allocations vs. RB Allocation for BS Rate 100 and 50: The former stems out 8 possible discrete rate allocations, while the latter generates 1 discrete resource allocation alone.

R	r_i	RB
100	11.57, 21.57, 33.58, 7.72, 10.36, 15.21	11, 21, 33, 8, 11, 16
100	11.57, 21.57, 33.58, 7.72, 10.36, 15.21	11, 21, 33, 8, 11, 15
100	11.57, 21.57, 33.58, 7.72, 10.36, 15.21	11, 21, 33, 8, 10, 16
100	11.57, 21.57, 33.58, 7.72, 10.36, 15.21	11, 21, 33, 8, 10, 15
100	11.57, 21.57, 33.58, 7.72, 10.36, 15.21	11, 21, 33, 7, 11, 16
100	11.57, 21.57, 33.58, 7.72, 10.36, 15.21	11, 21, 33, 7, 11, 16
100	11.57, 21.57, 33.58, 7.72, 10.36, 15.21	11, 21, 33, 7, 10, 16
100	11.57, 21.57, 33.58, 7.72, 10.36, 15.21	11, 21, 33, 7, 10, 15
50	10.46, 20.46, 24.21, 9.13, 9.13, 9.13	10, 20, 17, 1, 1, 1

optimization constraint. Excluding the RB sequences that violate the resource constraint of the relaxed continuous optimization improves the computational complexity significantly. Assuming n RB possibilities to be assigned to each UE, the computational complexity for M UEs decreases from $O(n^M)$ to at most $O(2^M)$. For a BS including 100 RB allocation possibilities and 100 UEs, this reduces the computations from 10^{200} to only 2^{100} as shown in Fig. 4.5. Besides, the algorithm is good. In the case where the BS bandwidth changed from 50 to 100 in steps of 1, the runtime was 16 s.

4.3.2   RB Allocation MATLAB Code

This code is a MATLAB implementation of the RB allocation solution algorithm simulations presented in Sect. 4.3.1.

```
1  %1. BS Algorithm.
2  %M=Number of UE under BS coverage.
3  %k=logarithmic utility parameters.
4  %a,b,c,d=Sigmodal utility parameters.
5  %eNRate=BS's rate
6  %r_UE=continuous UE rates.
7  %p=shadow price.
8  %w=bids
9  %-----------------------------
10  %Initialization
11  M=6;
12  k=[15 3 0.5];
13  a=[3 3 1];
14  b=[10 20 30];
15  c=(1+exp(a.*b))./(exp(a.*b));
16  d= 1./(1+exp(a.*b));
```

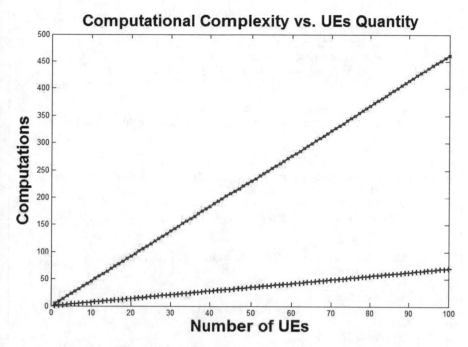

Fig. 4.5 Computational Complexity: The red curve depicts the semilog computational complexity vs. the UE quantity with non-violating sequence possibilities. The blue curve is the semilog computational complexity without the violating sequences

```
17   eNRate=50:100;
18   r_UE=1:100;
19   rM=max(r_UE);
20   MxItr=40;
21   r_inflection = [b zeros(1,size(a,2))];
22   r_min=zeros(1,M);
23   %UE Utility vs. UE continuous rate.
24   [SgUt,LgUt]=UeUtFn(k,a,b,c,d,r_UE,rM);
25   %Calculate continuous UE rates, shadow price, and bids.
26   for RtIn=1:length(eNRate)
27       %Initial Bids
28       w=10*ones(1,M);
29       n=1;
30       while (n<MxItr)
31           w_old=w;
32           p(n)=eNB_Rt(w,eNRate(RtIn));
33           for i=1:M
34               r_opt(i)=max(fzero(@(x) ...
                     utility(x,i,p(n),k,a,b,d),[.01 1000]),r_min(i));
35               w(i)=r_opt(i)*p(n);
36               if abs(w_old(i)-w(i))>(5.*exp(-0.1*n))
37                   w(i)=w_old(i)+(5.*exp(-0.1*n)).*sign(w(i)
38                       -w_old(i));
```

```
39              end
40          end
41          Sp(RtIn,n)=p(n);
42          w_sim{RtIn}(n,:)=w;
43          r_sim{RtIn}(n,:)=w./p(n);
44          n=n+1;
45      end
46      %Calculate resource blocks.
47      RsBl{RtIn}=DscrtRte(r_sim{RtIn}(end,:),eNRate(RtIn));
48  end
49  %Plots
50  n=n-1;
51  UtPl(r_UE,SgUt,LgUt,n,r_sim,w_sim,Sp,eNRate,RsBl);
52  %2.UE Utility vs. UE continuous rate.
53  %k=logarithmic utility parameters.
54  %c,d=Sigmodal utility parameters.
55  %r_UE=continuous UE rates.
56  %----------------------------------
57  function [SgUt,LgUt]=UeUtFn(k,a,b,c,d,r_UE,rM)
58      format long;
59      syms x
60      for i=1:length(c)
61          SgUt(i,:)=c(i).*(1./(1+exp(-a(i).*(x-b(i))))-d(i));
62      end
63      for i=1:length(k)
64          LgUt(i,:)=log(k(i).*x+1)./(log(k(i).*rM+1));
65      end
66      SgUt=subs(SgUt,x,r_UE);
67      LgUt=subs(LgUt,x,r_UE);
68  end
69  %3. Shadow price calculation.
70  %w=bids
71  %eNRate=BS's rate
72  %p=shadow price.
73  %----------------------------------
74  function p=eNodeB(w,eNRate)
75      p=sum(w)/eNRate;
76  end
77  %4. Get continuous rates.
78  %k=logarithmic utility parameters.
79  %a,b,c,d=Sigmodal utility parameters.
80  %r_UE=continuous UE rates.
81  %----------------------------------
82  function f=utility(x,i,p,k,a,b,d)
83      for j=1:length(a)
84          m(j)=exp(-a(j).*(x-b(j)));
85          dy_sig(j)=a(j).*m(j)./((1+m(j)).*(1-d(j).*(1+m(j))));
86          dy_log(j+length(a))=k(j)./((1+k(j).*x).*log(1+k(j).
87          *x));
88          dy(j)=dy_sig(j);
89          dy(j+length(a))=dy_log(j+length(a));
90      end
91      f=dy(i)-p;
92  end
```

```
93   %5. Discrete Rate Allocations.
94   %DsUp=Ceiling of continuous rates
95   %DsDn=Roof of continuous rates
96   %BndPnt=Possible ceiling and roof per continuous rate.
97   %DsCmb=Possible discrete rates.
98   %VlDsCmb=Valid discrete candidates.
99   %VlRt=Valid discrete rate indices.
100  %RsBl=Resource blocks possible which give maximum utility.
101  %-----------------------------------
102  function RsBl=DscrtRte(CntRt,eNR)
103      DsUp=ceil(CntRt);
104      DsDn=floor(CntRt);
105      DsDn(find(DsDn==0))=1;
106      BndPnt=[DsUp;DsDn];
107      C=mat2cell(BndPnt,size(BndPnt,1),ones(size(BndPnt,2),1));
108      DsCmb=CmbFn(C{:});
109      RtSum=sum(DsCmb,2);
110      VlRt=find(sum(RtSum,2)<=eNR);
111      VlDsCmb=DsCmb(VlRt,:);
112      PrpFr=UtCl(VlDsCmb);
113      RsBl=VlDsCmb(find(PrpFr==max(PrpFr)),:);
114  end
115  %6. Discrete allocation possibilities.
116  %varargin=Input vectors  whose combinations are sought.
117  %Cmb=Combinations of input vectors.
118  %-----------------------------------------
119  function Cmb = CmbFn(varargin)
120      error(nargchk(1,Inf,nargin)) ;
121      q =~cellfun('isempty',varargin) ;
122      if any(~q),
123          warning('Empty inputs result in an empty ...
                  output.') ;
124          Cmb = zeros(0,nargin) ;
125      else
126          ni = sum(q) ;
127          argn = varargin{end} ;
128          if ischar(argn) && (strcmpi(argn,'matlab') || ...
                  strcmpi(argn,'john')),
129              ni = ni-1 ;
130              ii = 1:ni ;
131              q(end) = 0 ;
132          else
133              ii = ni:-1:1 ;
134          end
135          if ni==0,
136              Cmb = [] ;
137          else
138              args = varargin(q) ;
139              if ni==1,
140                  Cmb = args{1}(:) ;
141              else
142                  [Cmb{ii}] = ndgrid(args{ii}) ;
143                  Cmb = reshape(cat(ni+1,Cmb{:}),[],ni) ;
144              end
```

```
145            end
146        end
147    end
148    %7. Calculate utility function.
149    %-----------------------------------
150    function PrpFr=UtCl(VlDsCmb)
151        k=[15 3 0.5];
152        a=[3 3 1];
153        b=[10 20 30];
154        c=(1+exp(a.*b))./(exp(a.*b));
155        d= 1./(1+exp(a.*b));
156        rM=100;
157        for j=1:size(VlDsCmb,1)
158            for i=1:length(a)
159                Utl(i) = c(i).*(1./(1+exp(-a(i).*(VlDsCmb(j,i)
160                    -b(i))))-d(i));
161            end
162            for i=1:length(k)
163                Utl(length(a)+i)=log(k(i).*VlDsCmb(length(a)+i)+1)
164                    ./(log(k(i).*rM+1));
165            end
166            PrpFr(j)=sum(abs(log(Utl)));
167            Utl=0;
168        end
169    end
170    %8.UE Utility plots
171    %r_UE=continuous UE rates.
172    %-----------------------------------
173    function UtPl(r_UE,SgUt,LgUt,n,rCn,wB,Sp,eNRate,RsBl)
174
175        figure(1);
176        plot(r_UE,SgUt);
177        hold on;
178        plot(r_UE,LgUt);
179        hold off;
180        title('UE Utility vs. Continuous Rates');
181        ylabel('\itU_i(r_i)');
182        xlabel('\itr_i');
183        %Plot logarithm of utilities vs. continuous rates.
184        figure(2);
185        plot(r_UE,log(SgUt));
186        hold on;
187        plot(r_UE,log(LgUt));
188        hold off;
189        title('Logarithm of UE Utility vs. Continuous Rates');
190        ylabel('\itlog(U_i(r_i))');
191        xlabel('\itr_i');
192        %Plot utilities vs. discrete rates.
193        figure(3);
194        for i=1:(length(r_UE)-1)
195            plot([r_UE(i) r_UE(i+1)],[SgUt(:,i) SgUt(:,i)]);
196            plot([r_UE(i) r_UE(i+1)],[LgUt(:,i) LgUt(:,i)]);
197            hold on;
198        end
```

```
199      plot([r_UE(end) r_UE(end)+1],[SgUt(:,end) SgUt(:,end)]);
200      plot([r_UE(end) r_UE(end)+1],[LgUt(:,end) LgUt(:,end)]);
201      hold off;
202      title('UE Utility vs. Resource Blocks');
203      ylabel('\itU_i(r_i)');
204      xlabel('\itr_i');
205      figure(4);
206      %Plot logarithm of utilities vs. continuous rates.
207      for i=1:(length(r_UE)-1)
208          plot([r_UE(i) r_UE(i+1)],[log(SgUt(:,i)) ...
                 log(SgUt(:,i))]);
209          plot([r_UE(i) r_UE(i+1)],[log(LgUt(:,i)) ...
                 log(LgUt(:,i))]);
210          hold on;
211      end
212      plot([r_UE(end) r_UE(end)+1],[log(SgUt(:,end)) ...
             log(SgUt(:,end))]);
213      plot([r_UE(end) r_UE(end)+1],[log(LgUt(:,end)) ...
             log(LgUt(:,end))]);
214      hold off;
215      title('Logarithm of UE Utility vs. Resource Blocks');
216      ylabel('\itlog(U_i(r_i))');
217      xlabel('\itr_i');
218      %Plot continuous rates allocated vs. iterations.
219      figure(5);
220      plot(1:n,rCn{1});
221      title('UE Allocated Continuous Rates vs. Iterations');
222      ylabel('\itr_i(n)');
223      xlabel('Iterations (\itn)');
224      %Plot UE bids vs. iterations.
225      figure(6);
226      plot(1:n,wB{1});
227      title('UE Bids vs. Iterations');
228      ylabel('\itw_i(n)');
229      xlabel('Iterations (\itn)');
230      %Plot shadow price vs. iterations.
231      figure(7);
232      plot(1:n,Sp(1,:));
233      title('Shadow price vs. Iterations');
234      ylabel('\itp(n)');
235      xlabel('Iterations (\itn)');
236      %Plot UE rates, UE bids, and shadow prices vs. BS rates.
237      UE_ER=[];
238      W_ER=[];
239      Sp_ER=[];
240      RsBl_ER=[];
241      for i=1:length(eNRate)
242          UE_ER=[UE_ER ; rCn{i}(end,:)];
243          W_ER=[W_ER ; wB{i}(end,:)];
244          Sp_ER=[Sp_ER ; Sp(i,end)];
245          RsBl_ER=[RsBl_ER ; RsBl{i}(1,:)];
246      end
247      figure(8);
248      plot(eNRate,UE_ER);
```

```
249       title('UE continuous rates vs. eNB Rate');
250       ylabel('\itr_i');
251       xlabel('\itR');
252       figure(9);
253       plot(eNRate,W_ER);
254       title('UE Bids vs. eNB Rate');
255       ylabel('\itw_i');
256       xlabel('\itR');
257       figure(10);
258       plot(eNRate,Sp_ER);
259       title('Shadow Price vs. eNB Rate');
260       ylabel('\itp');
261       xlabel('\itR');
262       figure(11);
263       plot(eNRate,RsBl_ER);
264       title('Resource Blocks vs. eNB Rate');
265       ylabel('Resource Block');
266       xlabel('\itR');
267   end
```

4.4   Chapter Summary

In this chapter, we investigated an RB allocation optimization with a proportional fairness policy to allocate resources to UEs by formulating the RB allocation as an integer optimization, hard to solve. To find a solution, we leveraged the Lagrangian relaxation to change the discrete RB allocation optimization to a continuous rate allocation convex optimization and solved it via the Lagrange multipliers of the dual problem to the continuous optimization. The relaxed continuous rate allocation solution is a series of rates given by the BS to the UEs. When continuous rates were assigned to the UEs, the original discrete RB optimization was approached by using the boundary discrete points right above and below the computed continuous rates. These values generate a series of candidate discrete rates, referred to as RBs, for the UEs with a small computational complexity. Besides, we evaluated the candidate RBs satisfying the constraints of the relaxed continuous rate allocation optimization, and constraint-satisfying ones are selected to locate those maximizing the proportional fairness utility functions. These RBs form a set of feasible discrete RBs for the UEs. The sensitivity of the solution algorithm to RB allocation to the variations in the available BS bandwidth was studied. We realized as the BS resources become more, both UE bid and the shadow prices decrease. In addition, we explained that the proposed optimization prioritized UEs with real-time applications. Furthermore, we explained that as the BS bandwidth exceeds the addition of inflection points in the relaxed continuous resource allocation optimization, it assigns more resources to UEs running delay-tolerant applications. At conclusion, we found out that the RB allocation algorithm is computationally efficient of the order of $O(2^M)$ vs. being of polynomial complexity.

References

1. H. Shajaiah, A. Abdelhadi, C. Clancy, Spectrum sharing between public safety and commercial users in 4G-LTE, in *IEEE International Conference on Computing, Networking and Communications (ICNC)*, 2014
2. H. Shajaiah, A. Abdelhadi, C. Clancy, Utility proportional fairness resource allocation with carrier aggregation in 4G-LTE, in *IEEE Military Communications Conference (MILCOM)*, 2013
3. M. Yokoo, E. Durfee, T. Ishida, K. Kuwabara, The distributed constraint satisfaction problem: formalization and algorithms. IEEE Trans. Knowl. Data Eng. **10**(5), 673–685 (1998)
4. M. Andrews, K. Kumaran, K. Ramanan, A. Stolyar, P. Whiting, R. Vijayakumar, Providing quality of service over a shared wireless link. IEEE Commun. Mag. **39**(2), 150–154 (2001)
5. M. Ghorbanzadeh, A. Abdelhadi, C. Clacy, *Cellular Communications Systems in Congested Environments Resource Allocation and End-to-End Quality of Service Solutions with MATLAB* (Springer, Berlin, 2017)
6. M. Ghorbanzadeh, Resource allocation and end-to-end quality of service for cellular communications systems in congested and contested environments. Ph.D. Thesis, Virginia Tech, 2015
7. H. Shajaiah, M. Ghorbanzadeh, A. Abdelhadi, C. Clancy, Application-aware resource allocation based on channel information for cellular networks, in *2019 IEEE Wireless Communications and Networking Conference (WCNC)*, pp. 1–6, 2019
8. M. Ghorbanzadeh, A. Abdelhadi, C. Clancy, Application-aware resource allocation of hybrid traffic in cellular networks. IEEE Trans. Cogn. Commun. Netw. **3**(2), 226–241 (2017)
9. J. Lee, R. Mazumdar, N. Shroff, Non-convex optimization and rate control for multi-class services in the internet. IEEE/ACM Trans. Netw. **13**(4), 827–840 (2005)
10. A. Abdelhadi, C. Clancy, A utility proportional fairness approach for resource allocation in 4G-LTE, in *IEEE International Conference on Computing, Networking, and Communications (ICNC), CNC Workshop*, 2014
11. A. Abdelhadi, C. Clancy, A robust optimal rate allocation algorithm and pricing policy for hybrid traffic in 4G-LTE, in *IEEE International Symposium on Personal, Indoor, and Mobile Radio Communications (PIMRC)*, 2013
12. A. Abdelhadi, C. Clancy, J. Mitola, A resource allocation algorithm for users with multiple applications in 4G-LTE, in *ACM Workshop on Cognitive Radio Architectures for Broadband (MobiCom Workshop CRAB)*, 2013
13. K. Kiwiel, T. Larsson, P. Lindberg, Lagrangian relaxation via ballstep subgradient methods. Math. Oper. Res. **32**(3), 669–686 (2007)
14. Y. Chen, M. Ghorbanzadeh, K. Ma, C. Clancy, R. McGwier, A hidden Markov model detection of malicious android applications at runtime, in *2014 23rd Wireless and Optical Communication Conference (WOCC)*, 2014
15. M. Ghorbanzadeh, E. Visotsky, P. Moorut, W. Yang, C. Clancy, Radar inband and out-of-band interference into LTE macro and small cell uplinks in the 3.5 GHz band, in *2015 IEEE Wireless Communications and Networking Conference (WCNC)*, 2015
16. M. Ghorbanzadeh, E. Visotsky, P. Moorut, W. Yang, C. Clancy, Radar in-band interference effects on macrocell LTE uplink deployments in the U.S. 3.5 GHz band, in *2015 International Conference on Computing, Networking and Communications (ICNC)*, 2015
17. M. Ghorbanzadeh, A. Abdelhadi, C. Clancy, A utility proportional fairness bandwidth allocation in radar-coexistent cellular networks, in *Military Communications Conference (MILCOM)*, 2014

Chapter 5
Resource Allocation with Channel

5.1 Introduction

In Chap. 3, we introduced a novel convex utility proportional fairness maximization for optimal resource allocation in wireless networks and outfitted the optimization with the subscriber, application status, and service differentiations parameterized, respectively, as UE subscription weights, application status weights, and application utility functions. Over there, we developed a centralized architecture for the proposed resource allocation which assigned application rates by the BS in a single stage in response to the application utility parameters sent by the UEs to the BSs. Moreover, we provided with a distributed architecture for the same radio resource allocation framework which was introduced in Chap. 3, which accounted for application types and temporal usages as well as UE priorities, and assigned application rates in two stages from the BSs to the UEs and by the UEs to the running applications. While we saw the efficacy of the proposed methodology in a real-world WiFi network in Chap. 3, in a realistic large scale wireless communications system, there are other factors to be accounted for. Radio waves undergo various propagation effects [1] including path loss, absorption by oxygen and water vapor [2–4], and diffraction loss, collectively referred to as the channel.

When said that a UE has a bad channel, it means that based on its current position with respect to its serving BS, the signals transmitted between the BS and the UE and vice versa suffer from severe path loss and possibly diffraction loss so that the throughput for the UE is much lower. Transmitting under bad condition leads to transmission loss and bit errors; therefore, in order to transmit under bad channel conditions, a lower order MCS is used. Lower coding MCS means that the lower number of bits can be transmitted with one symbol. This chapter shows the inefficacy of the resource allocation methods introduced in Chaps. 3 and 4 under various channel conditions, it develops a distributed channel-aware architecture for the proposed resource allocation framework, it proves that the new channel-aware proposed mechanism is convex, it provides with solution algorithms for the

© The Author(s), under exclusive license to Springer Nature Switzerland AG 2022
M. Ghorbanzadeh, A. Abdelhadi, *Practical Channel-Aware Resource Allocation*,
https://doi.org/10.1007/978-3-030-73632-3_5

channel-aware resource allocation, it provides with simulation results under isolated scenarios to show the effectiveness of the proposed method in resource allocation with channel considerations, and it shows the results of the resource allocations under the abundance as well as scarcity of resources with channel effects considered.

The remainder of this chapter is organized as follows. Section 5.2 provides the motivation for including the channel in the resource allocation and its necessity. Section 5.3 discusses radio resource management in Long Term Evolution (LTE) systems. Section 5.4 discusses the inefficiencies that will come to happen by not considering the channel effect. Section 5.5 presents a channel-aware distributed architecture for the resource allocation framework that was introduced in Chap. 4. Section 5.6 provides with the solution algorithms for the resource allocation. Section 5.7 provides with the simulation results; and, Sect. 5.8 concludes and summarizes the current chapter.

5.2 Channel Quality, Modulation, and Coding

Channel coding, as a pillar in digital communication systems, can be considered as one of the differences between analog and digital systems, and it makes error detection/correction feasible. Error correction can be in the form of an Automatic Repeat Request (ARQ), where RX requests a retransmission of data in case an error is detected. Another error correction mechanism is Forward Error Correction (FEC), where redundant bits are added to the data bits under a block or convolutional coding [5]. Another mechanism is Turbo coding [1, 6–8], whose performance is within a few tenth of a dB from the Shannons limit [9–11].

Link adaptation is another feature of modern wireless systems and is a mechanism to match transmission parameters to the channel automatically. LTE link adaptation leverages the Adaptive Modulation and Coding (AMC), in which if the SINR is sufficiently high, higher order modulation schemes with higher spectral efficiency are used. This leads to higher bit rates, whereas a lower order modulation scheme, which is essentially more robust to transmission errors, yields in lower spectral efficiency. On the other hand, for a given modulation scheme, an appropriate code rate can be chosen depending on the channel quality [9, 12–16]. A better channel quality allows for a higher code rate, which in turn results in a higher data rate. LTE does this by means of a rate matching module (RMM) after the Turbo encoder to permit choosing proper code rates through puncturing and repetition [1]. Figure 5.1 shows the signal generation chain of an LTE PHY with Turbo coding and modulation modules.

As we mentioned, the channel quality is an important aspect of wireless systems. In LTE, the quality of DL channel is measured in the UE for the reference symbols [17] and transmitted to the Evolved Node-B (eNB) as CQI . CQI depends on the channel, noise, interference, and RX quality (e.g. analog front-end noise figure (NF) and digital signal processing (DSP) algorithm performance). That means a receiver with better front end or more powerful signal processing algorithms delivers a higher

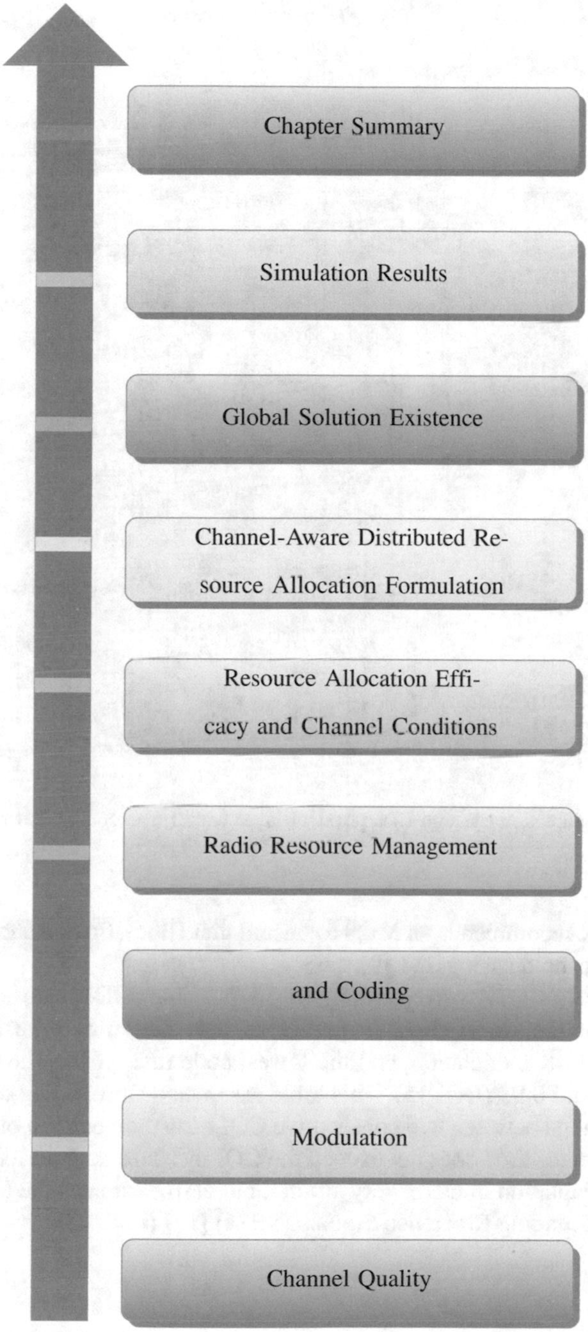

Chapter Summary

Simulation Results

Global Solution Existence

Channel-Aware Distributed Resource Allocation Formulation

Resource Allocation Efficacy and Channel Conditions

Radio Resource Management

and Coding

Modulation

Channel Quality

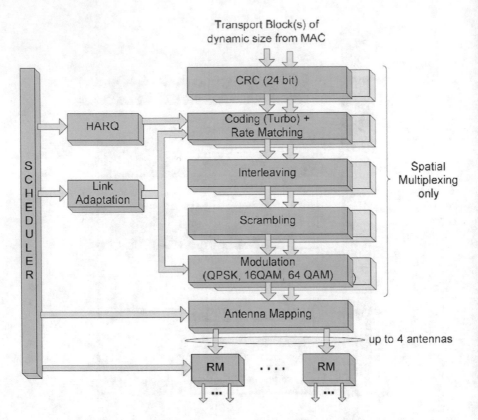

Fig. 5.1 Link Adaptation System: LTE physical layer with Turbo coding and modulation modules [9]

CQI. The SNR recommends an MCS to ensure that Block Error Rate (BLER) [5] is less than a threshold such as 0.1 [9].

In Table 5.1, CQIs corresponding to 16 possible MCSs are shown. CQI 1 is selected for the worst channel quality as it is the most robust transmission parameters, QPSK modulation, and the lowest code rate 0.076. The highest MCSs are 64 QAM and 0.93 (CQI 15). This table assumes a slow fading channel so that it does not change between two consecutive CQI measurements; in other words, the channel's coherent time does not exceed the CQI measurement period.

The link adaptation in UL is very similar, but eNB estimates the channel quality by using the Sounding Reference Signals (SRSs) [17] (Fig. 5.2).

Table 5.1 LTE DL link adaptation

MCS, code rate, coding efficiency			
0	No Transmission	–	–
1	$QPSK$	0.76	0.1523
2	$QPSK$	0.12	0.2344
3	$QPSK$	0.19	0.377
4	$QPSK$	0.3	0.6016
5	$QPSK$	0.44	0.877
6	$QPSK$	0.59	1.1758
7	$16QAM$	0.37	1.4766
8	$16QAM$	0.48	1.9141
9	$16QAM$	0.6	2.4063
10	$64QAM$	0.45	2.7305
11	$64QAM$	0.55	3.3223
12	$64QAM$	0.65	3.9023
13	$64QAM$	0.75	4.5234
14	$64QAM$	0.85	5.1152
15	$64QAM$	0.95	5.554

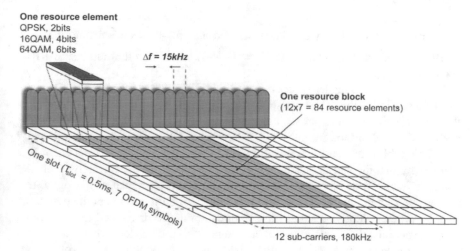

Fig. 5.2 LTE radio resource block [9]

5.3 Radio Resource Management

In LTE at the PHY layer, resources are managed using Resource Management (RM) modules that assign data blocks to RRBs. One RRB includes 12 subcarriers and one time slot. The resource management in LTE can be seen in Fig. 5.4, where CQI values are leveraged in choosing the RRBs. CQIs can be periodic where the Physical Uplink Control Channel (PUCCH) reports them, or Physical Uplink Shared Channel (PUSCH) reports aperiodic CQIs. The latter is used if BS wants the channel quality

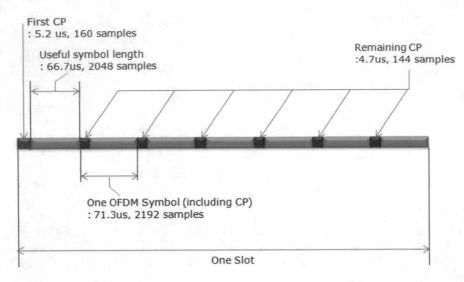

Fig. 5.3 LTE frame structure [18]

at a particular time. The bandwidth is divided into sub-bands consisting of RRBs, where the number of sub-bands is $N = \lceil \frac{N_{RB}^{DL}}{k} \rceil$, where k is the number of RRBs (Fig. 5.4).

The LTE has two modes of operation, FDD and TDD [1]. For the FDD mode, we have the following. Time duration for one frame is 10 ms, which means that there are 100 radio frames per second. There are 307,200 samples in the 10 ms LTE frame, i.e. 30.72 million samples in 1 s. There are 10 subframes in a frame and 2 slots in 1 subframe. Thus, a frame consists of 20 slots that are 0.5 ms each. It is noteworthy that the slot is not the smallest time unit in LTE. Normally, each slot contains 7 small time blocks called symbols, a certain time span of the signal that carries one spot in the I/Q constellation diagram [5]. Furthermore, at the beginning of each symbol, there is a small time span called cyclic prefix and the remaining is the symbol data. The cyclic prefix can be normal or extended, where the former leads to 7 data symbols and the latter yields in 6 data symbols. The explanations above are summarized in Fig. 5.3.

For the DL, the symbols are OFDM, and for the UL, they are SC-OFDMA [1]. The first symbol is a little bit longer than the others. Since there are 30.072 million samples per second, there would be 2048 bins/IFFT. Since the spacing between the subcarriers is identical (15 kHz), changing the system bandwidth generates different numbers of subcarriers as we can see in Table 5.2. As such, for 5, 10, and 20 MHz LTE, there are, respectively, 300, 600, and 1200 subcarriers. The symbols are 66.7 ms, and each RRB contains 7 symbols, because it is 1 time slot, and 12 subcarriers, which amounts to 84 resource elements in an RRB. Thus, 5, 10, and 20 MHz LTEs contain 25, 50, and 100 RBs in that order. The details about TDD LTE can be found

Fig. 5.4 System model for resource allocation

Table 5.2 LTE DL link adaptation

	Number of RRBs, number of IFFT	
1.4	6	128
3.0	15	256
5.0	25	512
10.0	50	1024
15.0	75	2048
20.0	100	2048

in [19]. While we present the formulations in this chapter general, our simulations are based on FDD due to simpler frame structure.

5.4   Resource Allocation Efficacy and Channel Conditions

The distributed architecture for optimal rate allocation in Chap. 3 was in accordance with Eq. (5.1), repeated here for the ease of reference, where for M UEs covered by an eNB in accordance with Fig. 5.4, $\mathbf{r} = [r_1, r_2, \ldots, r_M]$ is the UE allocated rate vector, R is the maximum available resources at the eNB, β_i is a subscription-dependent weight for the ith UE, and $V_i(r_i) = \prod_{j=1}^{N_i} U_{ij}^{\alpha_{ij}}(r_{ij})$ is the ith UE aggregate utility function.

$$\max_{\mathbf{r}} \quad \prod_{i=1}^{M} V_i^{\beta_i}(r_i)$$

$$\text{subject to} \quad \sum_{i=1}^{M} r_i \leq R, r_i \geq 0, \quad i = 1, 2, \ldots, M \tag{5.1}$$

And, the application rates were allocated using Eq. (5.2), where $\mathbf{r}_i = [r_{i1}, r_{i2}, \ldots, r_{iN_i}]$ is the application rate allocation vector such that its jth component indicates the bandwidth allotted by the ith UE to its jth application and r_i^{opt} is the ith UE rate allocated by eNB.

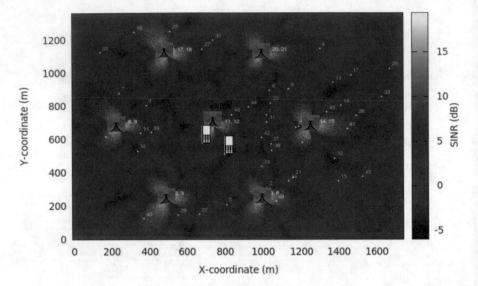

Fig. 5.5 System model and channel for resource allocation

$$\max_{\mathbf{r}_i} \quad \prod_{j=1}^{N_i} U_{ij}^{\alpha_{ij}}(r_{ij})$$

$$\text{subject to} \quad \sum_{j=1}^{N_i} r_{ij} \le r_i^{\text{opt}}, r_{ij} \ge 0, \qquad j = 1, 2, \ldots, N_i \tag{5.2}$$

However, the model specified by Eqs. (5.2) and (5.2) does not account for the channel conditions of the UEs as discussed in Sect. 5.2. This is shown in Fig. 5.5, from which we see that eNB A, which is serving 2 of its UEs depicted, is having a bad channel to the UE in the blue area of the graph, which represents low SNRs from the heat map index on the right side, whereas the other UE is observing an excellent channel by having a high SNR.

Applying the algorithm from Eqs. (5.2) and (5.2) does not differentiate between these two users at all since it only looks at UE subscriptions, instantaneous app usages, and application types. Therefore, two identical applications in terms of QoS requirements for two identical UEs in terms of subscription type and application focus will have identical resources allocated to them solely based on the bit rate requirements of the application and the UE and application weights. However, in practice, the throughput (in terms of bits per second) for the UE with bad channel condition might not be met since this UE has a bad channel and it has to use a much lower order of MCS in order to have a low BLER. In fact, for this situation, the UE with bad channel needs far more RRBs in order to have its QoS requirements met.

5.5   Channel-Aware Distributed Resource Allocation Formulation

Here, we define the number of bits that we can transmit with k_i resource elements at the signal-to-noise ratio SNR_i as function f shown in Eq. (5.3). The number of bits transmittable with k_i resource elements is k_i times the number of bits that can be transmitted with only 1 resource element under the same SNR. On the other hand, f_1 is upper bounded according to the Shannon law such that we can write Eq. (5.6), where β is the resource block bandwidth.

$$r_i = f(k_i, SNR_i) = k_i \underbrace{f(1, SNR_i)}_{f_1} \tag{5.3}$$

Then, we can multiply by a proportionality factor to make it into an equation. In order to do this, we know that there are control and pilot per resource block. So, if we assume that there are m non-data control and pilot bits of total n resource elements available, then $\frac{n-m}{n}$ comes into the equation. Furthermore, if the frame time is T_f and the resource block time is T_{RB}, then $\frac{T_{RB}}{T_f}$ comes to the equation as well.

$$r_i \propto k_i \beta log_2(1 + SNR_i) \tag{5.4}$$

Therefore, we can write the following equation:

$$r_i = \beta k_i \frac{T_{RB}}{T_f} \frac{n-m}{n} \epsilon_i log_2(1 + SNR_i) \tag{5.5}$$

As a case in point, if there were 4 pilots in a resource block and 84 (7 symbols and 12 subcarriers) resource elements in a resource block, and subcarriers were 15 kHz so that the resource block was 180 kHz, considering 0.5 ms time slot (resource block time that consists of 7 symbols each 71.4 μ s.) leads to the following equation that is in kbps:

$$r_i = 8.5713 k_i \epsilon_i log_2(1 + SNR_i) \tag{5.6}$$

Then, the resource allocation in Eq. (5.1) can be written as Eq. (5.8), where $\mathbf{k}_i = [k_1, r_2, \ldots, r_M]$ is the resource block vector whose ith element is the number of resource blocks allocated to the ith UE, which is under SNR_i channel condition. The constraint in the optimization says that the total number of resource blocks should be less than or equal to the total number of data resource blocks which are available at the eNB and that the number of allocated resource blocks must be positive. We can use $\geq$ instead of $>$ in the latter condition, since nonzero allocation is implied by the multiplication of utilities (Chap. 3).

$$\max_{\mathbf{k}} \quad \prod_{i=1}^{M} V_i^{\beta_i} (\beta k_i \frac{T_{\mathrm{RB}}}{T_f} \frac{n-m}{n} \epsilon_i log_2(1 + SNR_i))$$

$$\text{subject to} \quad \sum_{i=1}^{M} k_i \le K, k_i \ge 0, \quad i = 1, 2, \dots, M$$

(5.7)

It is noteworthy that to apply the proposed algorithms to a realistic scenario, one must obtain the values for the parameters K and epsilon. The latter was added to change Eq. (5.6) to (5.5). It implies that the rate should be bounded by the Shannon capacity, and therefore it should be a function of code efficiency and SNR. Furthermore, the parameter K should be obtained by reducing the total number of resource blocks available minus the number of control, pilot, and synchronization channel.

5.5.1   Determining ϵ

The minimum SNRs required for different CQIs reported are given in the 3GPP documentation as in Table 5.3. The first four columns are excerpted from Table 5.1, and the columns 5–9 show the minimum SNRs required for various MIMO/HARQ modes at various CQIs [17]. Then, we select the minimum SNR required, and the linear scale of this minimum SNR is depicted in the last column. Then, the parameter ϵ will be as in Eq. (5.8).

$$\epsilon_i = \frac{\gamma_i}{log_2(1 + SNR_{i,\mathrm{lin}})}$$

(5.8)

In order to have a continuous graph, we curve fit the values of this parameter ϵ vs. SNR based on linear regression, which yields in Eq. (5.9), depicted in Fig. 5.6.

$$\epsilon_i = 0.0087 SNR_{\mathrm{dB}} + 0.6821$$

(5.9)

Therefore, Eq. (5.8) can be written as

$$\max_{\mathbf{k}} \quad \prod_{i=1}^{M} V_i^{\beta_i} (\beta k_i \frac{T_{\mathrm{RB}}}{T_f} \frac{n-m}{n} (0.0087 SNR_i + 0.6821) log_2(1 + SNR_i))$$

$$\text{subject to} \quad \sum_{i=1}^{M} k_i \le K, k_i \ge 0, \quad i = 1, 2, \dots, M$$

(5.10)

Table 5.3 Characterizing epsilon for different channels (M modulation, CR coding rate, SE spectral efficiency, T transmit mode [17])

CQI	M	CR	SE	T1	T2	T3	T4	T5	$SNR_{min,lin}$
0	NT	–	–	–	–	–	–	–	–
1	$QPSK$	0.76	0.1523	1.95	2.00	−7.00	−3.10	−4.80	0.199526231
2	$QPSK$	0.12	0.2344	4.00	4.05	−5.00	−1.15	−2.60	0.316227766
3	$QPSK$	0.19	0.377	6.00	5.10	−3.15	1.50	0.00	0.484172368
4	$QPSK$	0.3	0.6016	8.00	8.00	−1.00	4.00	2.60	0.794328235
5	$QPSK$	0.44	0.877	10.00	10.00	1.00	6.00	4.95	1.258925412
6	$QPSK$	0.59	1.1758	11.95	11.80	3.00	8.90	7.60	1.995262315
7	$16QAM$	0.37	1.4766	14.05	13.90	5.00	12.70	10.60	3.16227766
8	$16QAM$	0.48	1.9141	16.00	16.10	6.90	14.90	12.95	4.897788194
9	$16QAM$	0.6	2.4063	17.90	17.45	8.90	17.50	15.40	7.762471166
10	$64QAM$	0.45	2.7305	19.9	19.50	10.85	20.50	18.10	12.16186001
11	$64QAM$	0.55	3.3223	21.5	21.50	12.60	22.45	20.05	18.19700859
12	$64QAM$	0.65	3.9023	23.45	23.10	14.35	23.20	22.00	27.22701308
13	$64QAM$	0.75	4.5234	25.00	24.90	16.15	24.90	24.55	41.20975191
14	$64QAM$	0.85	5.1152	27.30	27.00	18.15	27.00	26.80	65.31305526
15	$64QAM$	0.95	5.554	22.00	29.10	20.00	29.10	29.60	100.00

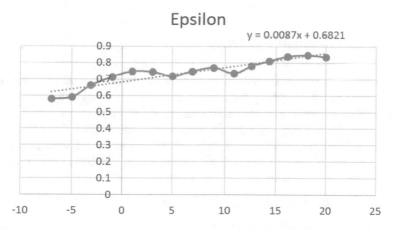

Fig. 5.6 Parameter ϵ vs. SNR

It is noteworthy that we assume that the SNR is ergodic and that the resource blocks are assigned within the coherence time/bandwidth so that we only have flat and slow fading.

Table 5.4 Characterizing K for different channels (*BW* bandwidth, *TB* transmit bandwidth, *TREPF* total resource element per frame, *TREPS* total resource element per second, *OREPF*, overhead resource element per second, *OREPS* overhead resource element per second, *DREPS* data resource element per second)

BW	TB	PUCCH	TREPF	TREPS	OREPF	OREPS	DREPS
10	9	2	84,000	84,000,000	16,857	67,143	67,143,000
20	18	2	168,000	168,000,000	27,454	140,546	140,546,000

5.5.2   Determining K

The number of available resource elements depends on the bandwidth for the LTE system and the mode of operation, i.e. FDD vs. TDD. The total number of resource elements in M resource blocks is $M \times 12 \times 7 \times 2 \times 10 \times 1000$, where 12 is the 12 subcarriers in 1 resource block, 7 is the number of symbols in 1 resource block, 2 is due to having 2 slots in a subframe, 10 is due to having 10 subframes in 1 frame, and 1000 is due to having 1000 frames in 1 s. However, there are data pilot and synchronization symbols that should be reduced from the total available to give out the number of data resource elements. This is depicted in Table 5.4.

5.6   Global Solution Existence

In order to follow the same procedure as in Chaps. 3 and 4, we have to prove that the logarithm of the UE utility function is convex. It is noteworthy that factoring the channel effect into the optimization changes the shape of the objective function under Eq. (5.10) from the original sigmoidal or logarithmic form. Doing so renders the optimization in Eq. (5.10) convex and therefore proves the existence of a globally optimal solution. In order to show the existence of such a globally optimal solution, first we present the following theorem as a steppingstone.

Lemma 5.6.1 *The objective function of the optimization in Eq. (5.10) is convex.*

Proof The sigmoidal utility function $U(r)$ from Chap. 3 is now replaced with $U(f(k, SNR))$ from Eq. (5.3). Therefore, for the sigmoidal utility, we have that

$$\frac{\partial f}{\partial k} = \underbrace{\frac{\partial f}{\partial U} \frac{\partial U}{\partial r}}_{\frac{\partial log(U)}{\partial r} > 0} \frac{\partial r}{\partial k} = \frac{\partial log(U)}{\partial r} (\beta k_i \frac{T_{RB}}{T_f} \frac{n-m}{n} \epsilon_i log_2(1 + SNR_i)) > 0$$

$$(5.11)$$

Now, we only should prove that the second derivative is negative. We have that

$$\left(\frac{\partial^2 log(U)}{\partial k^2} = \underbrace{\frac{\partial}{\partial k}\left(\frac{\partial log(U)}{\partial r}\right)\frac{\partial r}{\partial k}}_{g} = \underbrace{\frac{\partial g}{\partial k}\beta k_i \frac{T_{RB}}{T_f}\frac{n-m}{n}\epsilon_i log_2(1+SNR_i)}_{.>0} + g\underbrace{\frac{\partial^2 r}{\partial k^2}}_{=0}\right.$$

$$(5.12)$$

So, the positiveness or negativeness depends on $\frac{\partial g}{\partial k}$.

$$\frac{\partial g}{\partial k} = \frac{\partial}{\partial k}\left(\frac{\partial log(U)}{\partial r}\right) = \frac{\partial}{\partial k}\left(\frac{ade^{-a(r-b)}}{1-d(1+e^{-a(r-b)})} + \frac{ae^{-a(r-b)}}{1+e^{-a(r-b)}}\right)$$

$$(5.13)$$

Taking $\beta k_i \frac{T_{RB}}{T_f}\frac{n-m}{n}\epsilon_i log_2(1+SNR_i) - b)$ as Ψ, the derivative of the first term becomes

$$\frac{\partial}{\partial k}\frac{ade^{-a(r-b)}}{1-d(1+e^{-a(r-b)})} = ade^{-a\Psi}1-d(1+e^{-a\Psi})$$

$$= \underbrace{-a^2 d\Psi}_{A} e^{-a\Psi}\left[1-d(1+e^{-a\Psi})\right]\frac{-Ae^{-a\Psi}(ade^{-a\Psi})}{\left[1-d(1+e^{-a\Psi})\right]^2}$$

$$-\frac{aAe^{(Bk} ba)\left[1 \quad d(1 \mid e^{-(Bk-ba)})\right] \mid de^{-(Bk-ba)}}{\left[1-d(1+e^{-(Bk-ba)})\right]^2}$$

$$= \frac{-Aae^{-(Bk-ba)(1-d)}}{\left[1-d(1+e^{-(Bk-ba)})\right]^2}$$

$$(5.14)$$

And the derivative of the second term becomes

$$\frac{\partial}{\partial k}\frac{ae^{-a(r-b)}}{1+e^{-a(r-b)}} = \frac{\partial}{\partial k}\frac{ae^{-aB}}{1+e^{-aB}}$$

$$= \frac{(-a^2\beta k_i \frac{T_{RB}}{T_f}\frac{n-m}{n}\epsilon_i log_2(1+SNR_i)e^{-aBk-ba})(1+e^{-a(Bk-b)})}{(1+e^{-a(Bk-ba)})^2}$$

$$= \frac{-aAe^{-(Bk-ba)}(1+e^{-(Bk-ba)}) + Aae^{-(Bk-ba)}e^{-(Bk-ba)}}{\left[1-d(1+e^{-(Bk-ba)})\right]^2}$$

$$= \frac{-aAe^{-(Bk-ba)}}{\left[1-d(1+e^{-(Bk-ba)})\right]^2}$$

$$(5.15)$$

Therefore, the derivative comes as the following equation:

$$
\frac{\partial g}{\partial k} = -\frac{Aae^{-(Bk-ba)(1-d)}}{\left[1 - d(1 + e^{-(Bk-ba)})\right]^2} - \frac{aAe^{-(Bk-ba)}}{\left[1 - d(1 + e^{-(Bk-ba)})\right]^2}
$$

$$
= \frac{d(1 + e^{-(Bk-ba)})^2(1 - d) + \left[1 - d(1 + e^{-(Bk-ba)})\right]^2}{(1 - d(1 + e^{-(Bk-ba)}))^2 d(1 + e^{-(Bk-ba)})^2}
$$

$$
- Aae^{-(Bk-ba)} = -\frac{aAd^2(1 + e^{-2(Bk-ba)})e^{-(Bk-ba)} + (2d + 1)Aae^{-(Bk-ba)}}{(1 - d(1 + e^{-(Bk-ba)}))^2 d(1 + e^{-(Bk-ba)})^2}
$$

$$
\tag{5.16}
$$

Thus, the second derivative is negative, and hence it is a concave function. The proof for the logarithmic utility is similar.

Theorem 5.6.2 *The optimization problem in Eq. (5.8) is convex and has a unique tractable global optimal solution.*

Proof We can form the equivalent log-EURA optimization (5.17) by taking the logarithm of (5.8)

$$
\max_{\mathbf{k}} \quad \sum_{i=1}^{M} \beta_i \log(V_i(\beta k_i \frac{T_{\mathrm{RB}}}{T_f} \frac{n-m}{n}(0.0087 SNR_i + 0.6821) log_2(1 + SNR_i)))
$$

$$
\text{subject to} \quad \sum_{i=1}^{M} r_i \leq K, k_i \geq 0, \quad i = 1, 2, \ldots, M
$$

$$
\tag{5.17}
$$

The aggregate utility concavity based on Lemma 5.6.1 concludes that the channel-aware log-EURA optimization is convex [20], which in turn proves the convexity of the EURA problem in Eq. (5.8) due to their objective function equivalence. And, there exists a unique tractable global optimal solution for a convex optimization in general [20] and for EURA in particular.

5.6.1   Solution for Channel-Aware EURA Optimization

The solution of the distributed architecture channel-aware resource allocation formulation is achieved through the Lagrangian of the dual problem of the EURA optimization. Similarly to [21, 22], we deploy the duality for convex optimization problems to solve them efficiently. What proceeds is such an application of the duality to EURA and IURA constituents of the distributed rate allocation problem.

$$
L(\mathbf{k}, p) = \sum_{i=1}^{M} \log(V_i(\beta k_i \frac{T_{\mathrm{RB}}}{T_f} \frac{n-m}{n}(0.0087 SNR_i + 0.6821)
$$

$$\times \log_2(1 + SNR_i))) - p\left(\sum_{i=1}^{M} k_i + z - K\right)$$

$$= \sum_{i=1}^{M}\left(\log(V_i(\beta k_i \frac{T_{RB}}{T_f}\frac{n-m}{n}(0.0087SNR_i + 0.6821)\right.$$

$$\times \log_2(1 + SNR_i))) - pk_i\Big) + p(K - z)$$

$$= \sum_{i=1}^{M} L_i(k_i, p) + p(K - z) \tag{5.18}$$

Here, $z_i \geq 0$ is the slack variable and p is the Lagrange multiplier or the shadow price (price per unit bandwidth for all the M channels). Therefore, the ith UE bid for bandwidth can be written as $w_i = pk_i$, where $\sum_{i=1}^{M} w_i = p\sum_{i=1}^{M} k_i$. The first term in Eq. (3.19) is separable in k_i, so we have

$$\max_{\mathbf{k}} \sum_{i=1}^{M} (\log(V_i(\beta k_i \frac{T_{RB}}{T_f}\frac{n-m}{n}(0.0087SNR_i + 0.6821)log_2(1 + SNR_i))) - pk_i) =$$

$$\sum_{i=1}^{M} \max_{k_i}(\log(V_i(\beta k_i \frac{T_{RB}}{T_f}\frac{n-m}{n}(0.0087SNR_i + 0.6821)log_2(1 + SNR_i))) - pk_i),$$

and the dual problem objective function can be written as Eq. (5.19).

$$D(p) = \max_{\mathbf{k}} L(\mathbf{k}, p)$$

$$= \sum_{i=1}^{M} \max_{k_i}\left(\log(V_i(\beta k_i \frac{T_{RB}}{T_f}\frac{n-m}{n}(0.0087SNR_i + 0.6821)\right.$$

$$\times \log_2(1 + SNR_i))) - pk_i\Big)$$

$$+ p(K - z) = \sum_{i=1}^{M} \max_{k_i}(L_i(k_i, p)) + p(K - z) \tag{5.19}$$

Thus, the dual problem is formulated as

$$\min_{p} \quad D(p)$$

$$\text{subject to} \quad p \geq 0 \tag{5.20}$$

Leveraging the method of Lagrange multipliers, we have

$$\frac{\partial D(p)}{\partial p} = K - \sum_{i=1}^{M} k_i - z = 0 \tag{5.21}$$

Substituting by $\sum_{i=1}^{M} w_i = p \sum_{i=1}^{M} k_i$, we have Eq. (5.22), minimized to $p = \frac{\sum_{i=1}^{M} w_i}{K}$ at $z = 0$, where $w_i = pk_i$ is transmitted by the ith UE to the eNB.

$$p = \frac{\sum_{i=1}^{M} w_i}{K - z} \tag{5.22}$$

As such, we divide the channel-aware log-EURA problem (5.17) into simpler optimizations at the eNB (eNB EURA problem) and UEs (UE EURA problem). These are, respectively, shown in Eqs. (5.24) and (5.23) whose solutions, guaranteeing the utility proportional fairness in Eq. (3.15), are summarized in Algorithms 13 and 12 in that order.

$$\max_{r_i} \quad \log V_i(r_i) - pr_i$$
$$\text{subject to} \quad p \geq 0, r_i \geq 0, \quad i = 1, 2, \ldots, M \tag{5.23}$$

During the execution of the aforesaid algorithms, starting with $w_i(0) = 0$, the ith UE transmits an initial bid $w_i(1)$ to the eNB, which in turn subtracts the latterly received bid $w_i(n)$ and the formerly received one $w_i(n-1)$ and ceases the procedure if the difference is less than a threshold δ; otherwise, it computes and sends a shadow price $p(n) = \frac{\sum_{i=1}^{M} w_i(n)}{R}$ to covered UEs. The ith UE extracts its rate $k_i(n)$ from the received $p(n)$ such that $\log V_i(\beta k_i \frac{T_{RB}}{T_f} \frac{n-m}{n} (0.0087 SNR_i + 0.6821) log_2(1 + SNR_i)) - p(n)k_i$ is maximized. The rate $k_i(n)$ is employed to estimate the new bid $w_i(n) = p(n)k_i(n)$, transmitted to the eNB. This routine repeats until the bid difference $|w_i(n) - w_i(n-1)|$ falls below the threshold δ.

$$\min_{p} \quad D(p)$$
$$\text{subject to} \quad p \geq 0. \tag{5.24}$$

The solution $k_i(n)$ of the ith UE EURA optimization can be written as Eq. (5.25), then Algorithm (12) solves Eq. (5.26). This solution algebraically is the Lagrange multiplier solution for Eq. (5.23) and geometrically is the intersection point of the horizontal line $y = p(n)$ with the curve given by Eq. (5.27).

$$k_i(n) = \arg\max_{k_i} \left(\log V_i(\beta k_i \frac{T_{RB}}{T_f} \frac{n-m}{n} (0.0087 SNR_i + 0.6821) \right.$$
$$\left. \times \log_2(1 + SNR_i)) - p(n)k_i \right) \tag{5.25}$$

Algorithm 12 UE channel-aware EURA optimization algorithm

Send initial bid $w_i(1)$ to eNB.
loop
  Receive shadow price $p(n)$ from eNB.
  if STOP from eNB **then**
    Calculate allocated rate $k_i^{\text{opt}} = \frac{w_i(n)}{p(n)}$.
    STOP
  else
    Solve $k_i(n) = \arg\max_{k_i}\left(\log V_i(\beta k_i \frac{T_{\text{RB}}}{T_f} \frac{n-m}{n}(0.0087 SNR_i + 0.6821)log_2(1 + SNR_i)) - \right.$
    $\left. p(n)k_i\right)$.
    Send new bid $w_i(n) = p(n)k_i(n)$ to eNB.
  end if
end loop

Algorithm 13 eNB EURA optimization algorithm

---

loop
  Receive bids $w_i(n)$ from UEs. {Let $w_i(0) = 1 \; \forall i$}
  if $|w_i(n) - w_i(n-1)| < \delta \; \forall i$ **then**
    Allocate rates, $k_i^{\text{opt}} = \frac{w_i(n)}{p(n)}$ to user i.
    STOP
  else
    Calculate $p(n) = \frac{\sum_{i=1}^{M} w_i(n)}{K}$.
    Send new shadow price $p(n)$ to all UEs.
  end if
end loop

$$\frac{\partial \log V_i(\beta k_i \frac{T_{\text{RB}}}{T_f} \frac{n-m}{n}(0.0087 SNR_i + 0.6821)log_2(1 + SNR_i))}{\partial k_i} = p(n)$$

(5.26)

$$y = \frac{\partial \log V_i(\beta k_i \frac{T_{\text{RB}}}{T_f} \frac{n-m}{n}(0.0087 SNR_i + 0.6821)log_2(1 + SNR_i))}{\partial k_i}$$

(5.27)

5.6.2 IURA Global Optimal Solution

The strictly increasing nature of the logarithm function means that the IURA objective function in Eq. (5.2), i.e., $\prod_{j=1}^{N_i} U_{ij}^{\alpha_{ij}}(r_{ij})$, corresponds to $\sum_{j=1}^{N_i} \alpha_{ij} \log(U_{ij}(r_{ij}))$. So Eq. (3.16) can be reformulated as Eq. (5.28), referred to as the log-IURA problem for which corollary (3.3.2) is conceivable. It is noteworthy that for Eq. (5.28), the value of r_i is given as Eq. (5.29).

Algorithm 14 UE IURA algorithm

---
loop
    Receive r_i^{opt} from eNB. {by EURA Algorithms}
    Solve
    $\mathbf{r}_i = \arg\max_{\mathbf{r}_i} \sum_{j=1}^{N_i} (\alpha_{ij} \log U_{ij}(r_{ij}) - p_I r_{ij}) + p_I r_i^{\text{opt}}$ {$\mathbf{r}_i = \{r_{i1}, r_{i2}, \ldots, r_{iN_i}\}$}
    Allocate r_{ij} to the jth application.
end loop

$$\max_{\mathbf{r}_i} \sum_{j=1}^{N_i} \alpha_{ij} \log U_{ij}(r_{ij})$$

$$\text{subject to} \sum_{i=1}^{N_i} r_{ij} \leq \beta k_i^{opt} \frac{T_{\text{RB}}}{T_f} \frac{n-m}{n} (0.0087 SNR_i + 0.6821)$$

$$\times \log_2(1 + SNR_i), r_{ij} \geq 0, \quad j = 1, 2, \ldots, N_i \tag{5.28}$$

$$r_i^{opt} = (\beta k_i^{opt} \frac{T_{\text{RB}}}{T_f} \frac{n-m}{n} (0.0087 SNR_i + 0.6821) log_2(1 + SNR_i)) \tag{5.29}$$

Corollary 5.6.3 *The IURA optimization problem in Eq. (5.2) is convex and has a unique tractable global optimal solution.*

Proof Substantiating Lemma 5.6.1 is concomitant with proving the concavity of the application utility functions' natural logarithm; this ascertains the convexity of the log-IURA problem in Eq. (5.28) [20]. Since log-IURA and IURA optimizations have equivalent objective functions, IURA optimization in Eq. (5.2) is also convex. Every convex optimization has a tractable global optimal solution in general [20] and so does the IURA optimization in particular.

Theorem 5.6.2 and Corollary 5.6.3 indicate that the distributed optimization in Sect. 5.4 is convex and it assigns rates optimally. The application rates r_{ij} are optimally assigned internally to the UEs in accordance with Algorithm 14, where the ith UE leverages the EURA allocated rate r_i^{opt} to solve $\mathbf{r}_i = \arg\max_{\mathbf{r}_i} \sum_{j=1}^{N_i} (\alpha_{ij} \log U_{ij}(r_{ij}) - p_I r_{ij}) + p_I r_{ij}^{\text{opt}}$.

5.7 Simulation Results

We consider 9 UEs and a BS serving the UEs. We assume a 10 MHz LTE system so that 67,143,000 resource elements are available. Then, we apply the algorithm to observe the rates allocated to the UEs and to the applications. We assume ergodic SNRs. Then, the throughput and the number of resources allocated are illustrated in Fig. 5.7a, b. Here, the bit rate requirements of the UEs were 0.25, 1, and 5 Mbps, and the SNRs were −5, 5, and 15 dB as we can see on the x-axis. As we can observe, for identical QoS requirements, more resources are allocated by the algorithm to the UEs at lower SNRs, i.e. worst channel quality. On the other hand, a smaller number of resources meet the same bit rate requirements as the lower SNR ones as a higher MCS can be used. Furthermore, the abundance of the resources allows the UEs to have their throughput met as we can see in Fig. 5.7a.

Next, we consider the same 9 UEs, but we put two applications in each UE and reduce resources to 5,000,000 resource elements. As we can see, more resources are allocated to the UEs with lower SNRs in order to meet their bit rate requirements (Fig. 5.7a). The throughput of the UEs are met as it is shown in Fig. 5.7b, and IURA algorithm distributed the resources to its applications. Since the applications are identical, resources are allocated between them equally, which is shown by the stack bar chart 5.8c.

Next, we reduce the resources to only 1,000,000 resource elements for the 9 UEs. As we can see from the throughput plot in Fig. 5.9a, the throughput of the UEs under band channel conditions is not met. This is in spite of the fact that more resources were allocated to these UEs according to Fig. 5.9b; however, there were simply not enough resources available to meet the QoS requirements of the applications running on these UEs.

Next, as depicted in Fig. 5.10, we reduce the number of resources to only 100,000. Under this circumstance, the network is severely suffering and the throughput of no application is met as there are no enough resources available.

5.7.1 MATLAB Code

Here is the MATLAB code used for this chapter.

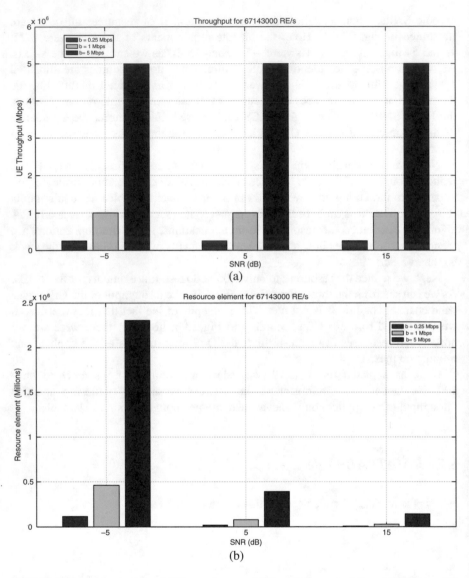

Fig. 5.7 The system contains 9 UEs, each concurrently running a real-time application with respective identically colored sigmoidal utility functions. The lower SNR UEs are allocated more resources so that their throughput meets their QoS requirements. (**a**) Throughput of the UEs. (**b**) Resources allocated to UEs

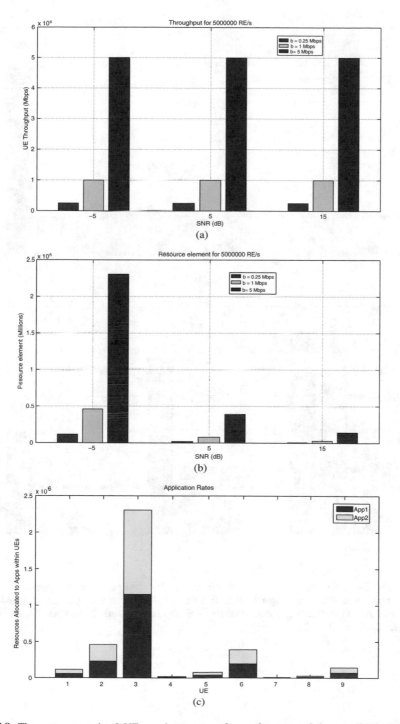

Fig. 5.8 The system contains 9 UEs, each concurrently running two real-time applications with respective identically colored sigmoidal utility functions. The lower SNR UEs are allocated more resources so that their throughput meets their QoS requirements. (**a**) Throughput of the UEs. (**b**) Resources allocated to UEs. (**c**) Resources allocated to UEs

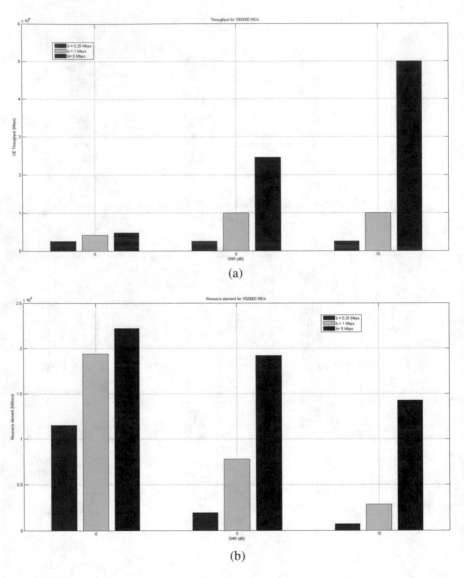

Fig. 5.9 The system contains 9 UEs, each concurrently running a real-time application with respective identically colored sigmoidal utility functions. The lower SNR UEs are allocated more resources so that their throughput meets their QoS requirements. (**a**) Throughput of the UEs. (**b**) Resources allocated to UEs

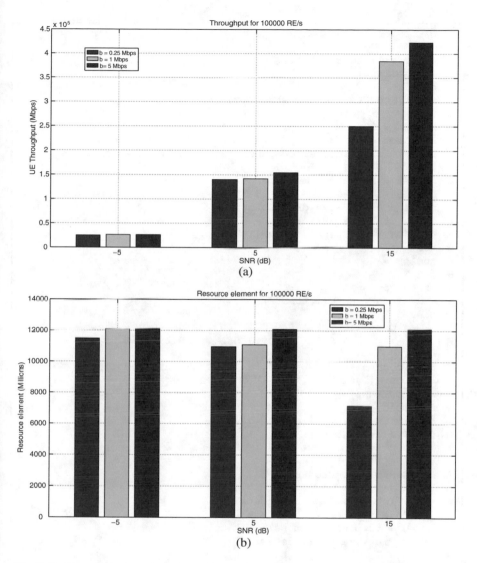

Fig. 5.10 The system contains 9 UEs, each concurrently running a real-time application with respective identically colored sigmoidal utility functions. The lower SNR UEs are allocated more resources so that their throughput meets their QoS requirements. (**a**) Throughput of the UEs. (**b**) Resources allocated to UEs

```
 1  %function ComputeUeApRateRe.m is the main file to call ...
        the code.
 2  close all;clear all;clc;
 3  format long;
 4  syms x
 5  global numUEs a b c d alpha coeffUE
 6  %LTE Parameters
 7  chanBW = 10;
 8  numPucch = 2;
 9  numDataRe = computeLteDataRePerSec(chanBW,numPucch);
10  %UE SNR
11  numUEs = 9;      % number of users
12  SNR_dB = [-5 -5 -5 5 5 5 15 15 15];
13  epsil = 0.0087*SNR_dB+0.6821;
14  SNR = 10.^(SNR_dB/10);
15  coeffUE = 8.57*epsil.*log2(1+SNR);
16  %UE Apps
17  numApp = 1;
18  %Intialize
19  if numApp == 1
20      alpha = ones(1,numUEs);
21  else
22      alpha = [0.1 0.5 0.9 0.1 0.5 0.9 0.8 0.4 0.2;...
23              0.9 0.5 0.1 0.9 0.5 0.1 0.2 0.6 0.8];
24  end
25  bOld = ones(1,numUEs).*[0.25 1 5 0.25 1 5 0.25 1 5]*(10^6);
26  %a = ones(1,numUEs).*[15 12 9 6 3 1 2 4 9];
27  a = coeffUE;
28  b = bOld./a;
29  c = (1+exp(a.*b))./(exp(a.*b));
30  d = 1./(1+exp(a.*b));
31  % k = ones(1,numUEs).*[15 12 9 6 3 1 8 4 2];
32  %UtilityPlot;
33  numDataRe = numDataRe;
34  %%%%%%%%% UE & eNodeB %%%%%%%%%%%%%%%%%%%
35  for i_rate = 1:length(numDataRe)
36      Rate(i_rate) = numDataRe;
37      w = 10*ones(1,numUEs);
38      r_min = zeros(1,numUEs);
39      Δ = 1;
40      time = 0;
41      while (Δ > 0.001)
42          time = time + 1;
43          w_old = w;
44          t(time) = time;
45          p(time) = eNodeB(w, Rate(i_rate));
46          for i = 1:numUEs
47              pp = p(time);
48              ii = i;
49              f = utility_UE(x,ii,pp);
50              %r0 = 5;
51              %g = fsolve(@myfun,r0);
```

```
52          %soln(i) = fzero(@(x) ...
                utility_UE(x,ii,pp),[.001 100]);
53          if (a(ii) <= pp)
54              dis('error: shadow price should be less ...
                    than a(ii).');
55          else
56              soln(i) = b(ii)+((log(a(ii)/pp - 1))/a(ii));
57          end
58          r_opt(i) = max(soln(i), r_min(i));
59          w(i) = r_opt(i)* p(time);
60          if abs(w_old(i)-w(i)) > (5.* ...
                exp(-0.1*time))%(10 ./ time)
61              w(i) = w_old(i) + (5.* exp(-0.1*time)) .* ...
                    sign(w(i)-w_old(i));
62          end
63       end
64       w_sim(time,:) = w;
65       r_sim(time,:) = w_sim(time,:)./p(time);
66       Δ = max(abs(w-w_old));
67    end
68    t_final =time;
69    r_ss(i_rate,:) = r_sim(time,:);
70    w_ss(i_rate,:) = w_sim(time,:);
71    p_ss(i_rate) = p(time);
72    time_ss(i_rate) = time;
73    time_ss2(i_rate) = 1;
74    throughput_UE(i_rate,:) = r_ss.*coeffUE;
75    %Some plots
76    figure;
77    bar([-5 5 15],[r_ss(1:3);r_ss(4:6);r_ss(7:9)])
78    grid on
79    xlabel('SNR (dB)')
80    ylabel('number of resource element units')
81    legend('b = 0.25','b = 1','b= 5');
82    title('Resources allocated 6MHz');
83    figure;
84    bar([-5 5 15],[throughput_UE(1:3);throughput_UE(4:6);
85     throughput_UE(7:9)])
86    grid on
87    xlabel('SNR (dB)')
88    ylabel('UE Throughput')
89    legend('b =0.25','b = 1','b= 5');
90    title('Throughput');
91    %%%%%%%%% INSIDE UE   %%%%%%%%%%%%%%%%%%%%
92    wUE = [10 10 10 10 10 10 10 10 10 10 10 10];
93    w_oldUE = [100 100 100 100 100 100 100 100 100 100 ...
                100 100];
94    r_minUE = [0 0 0 0 0 0 0 0 0 0 0 0]; % initial new ...
                min_rate of optimization
95    ΔUE = 1;
96    for i = 1: NoUE
97        time = 0;    % initate for each UE
98        while  (time<40)%(ΔUE > 0.01) %
99            time = time + 1;
```

```
100               t(time) = time;
101               ii =i;
102               ii2 =i + NoUE;
103               w_oldUE(ii) = wUE(ii);
104               w_oldUE(ii2) = wUE(ii2);
105               R2 = r_sim(t_final, ii);        % solution ...
                     from algorithm1 so it is the maximum for ...
                     2nd algorithm
106               pUE(time,ii) = UE(ii, wUE, R2);  % solve ...
                     centralized on the UE
107               pUE(time,ii2) = pUE(time,ii);
108               %ppUE(time,ii) = pUE(time);
109               ppUE(ii) = pUE(time,ii);
110               soln(ii) = fzero(@(x) ...
                     utility_UE(x,ii,ppUE(ii)),[.0001 100]);
111               r_optUE(ii) = max(soln(ii), r_minUE(ii));
112               wUE(ii) = r_optUE(ii) * ppUE(ii);
113               soln(ii2) = fzero(@(x) ...
                     utility_UE(x,ii2,ppUE(ii)),[.0001 100]);
114               r_optUE(ii2) = max(soln(ii2), r_minUE(ii2));
115               wUE(ii2) = r_optUE(ii2) * ppUE(ii);
116               if abs(w_oldUE(ii)-wUE(ii)) > (5.* ...
                     exp(-0.1*time))%(10 ./ time)
117                   wUE(ii) = w_oldUE(ii) + (5.* ...
                         exp(-0.1*time)) .* ...
                         sign(wUE(ii)-w_oldUE(ii));
118                   %1
119               end
120   %              if (r_optUE(ii) > r_inflectionUE(ii) && ...
          r_inflectionUE(ii) > 0)
121   %                  r_minUE(ii) = r_inflectionUE(ii);
122   %             end
123
124               if abs(w_oldUE(ii2)-wUE(ii2)) > (5.* ...
                     exp(-0.1*time))%(10 ./ time)
125                   wUE(ii2) = w_oldUE(ii2) + (5.* ...
                         exp(-0.1*time)) .* ...
                         sign(wUE(ii2)-w_oldUE(ii2));
126                   %1
127               end
128   %              if (r_optUE(ii2) > r_inflectionUE(ii2) && ...
          r_inflectionUE(ii2) > 0)
129   %                  r_minUE(ii2) = r_inflectionUE(ii2);
130   %             end
131               w_simUE(time,ii) = wUE(ii);
132               w_simUE(time,ii2) = wUE(ii2);
133               r_simUE(time,ii) = ...
                     w_simUE(time,ii)./pUE(time,ii);
134               r_simUE(time,ii2) = ...
                     w_simUE(time,ii2)./pUE(time,ii2);
135               ΔUE = max(abs(wUE-w_oldUE));
136           end
137       end
138       r_ssUE(i_rate,:) = r_simUE(time,:);
```

```
139        w_ssUE(i_rate,:) = w_simUE(time,:);
140        %p_ss(i_rate) = p(time);
141        %time_ss(i_rate) = time;
142   end % end for i_rate
143   plot(Rate,r_ss)
144   xlabel('Rate');
145   legend('UE1','UE2','UE3','UE4','UE5','UE6');
146   ylabel('UE rates')
147   figure;
148   plot(Rate,w_ss)
149   xlabel('Rate');
150   legend('UE1','UE2','UE3','UE4','UE5','UE6');
151   ylabel('UE bids ')
152   figure;
153   plot(Rate,time_ss,Rate,time_ss2)
154   xlabel('Rate');
155   ylabel('Iterations')
156   figure;
157   plot(Rate,r_ssUE)
158   xlabel('Rate');
159   legend('UE1 App1','UE1 App1','UE2 App1','UE2 App1','UE3 ...
             App1','UE3 App1');
160   ylabel('App rates')
161   figure;
162   plot(Rate,w_ssUE)
163   xlabel('Rate');
164   legend('UE1 App1','UE1 App1','UE2 App1','UE2 App1','UE3 ...
             App1','UE3 App1');
165   ylabel('App bids')
166   %figure;
167   %plot(Rate,time_ssUE)
```

This code *ComputeUeApRateRe.m* above calls computeLteDataRePerSec.m, which returns the number of LTE resource elements (REs) for a 10 MHz and a 20 MHz. The quantity of REs, used for the resource allocation, should include only data channels of LTE and not the control channels.

```
1   %computeLteDataRePerSec.m
2   function numDataRe = computeLteDataRePerSec(chanBW,numPucch)
3   if chanBW == 10
4       switch numPucch
5           case 2
6               numDataRe = 67143e3;
7           case 4
8               numDataRe = 63783e3;
9           case 6
10              numDataRe = 60423e3;
11          otherwise
12              disp('PUCCH can only be 2, 4, and 6');
13      end
14  elseif chanBW == 20
15      switch numPucch
```

```
16          case 2
17              numDataRe = 140564e3;
18          case 4
19              numDataRe = 137095e3;
20          case 6
21              numDataRe = 133735e3;
22          otherwise
23              disp('PUCCH can only be 2, 4, and 6');
24      end
25  else
26      error('LTE Channel Bandwidth is 10 or 20 MHz');
27  end
```

The code *ComputeUeApRateRe.m* above also calls the *eNodeB.m* code below to return the shadow price.

```
1  %eNodeB.m
2  function [p] = eNodeB(w,Rate)
3  R = Rate;
4  p = sum(w)/R;
```

Moreover, the code *ComputeUeApRateRe.m* above also calls the `utility_UE.m` code.

```
1  %utility_UE.m
2  function f = utility_UE(x,i,pp)
3  global  a b d alpha coeffUE
4  %y_sig(i) = log(c(i).*(1./(1+exp(-a(i).*(x-b(i))))-d(i)));
5  %y_log(i) = log(log(k(i).*x+1)./(1+ log(k(i).*100+1)));
6  m = exp(-a(i).*(x-b(i)));
7  dy_sig = (alpha(1,i).*a(i).*m)./((1+m).*(1-(d(i).*(1+m))));
8  dy = dy_sig;
9  f = dy-pp;
```

5.8   Chapter Summary

In this chapter, we developed channel-aware resource allocation for the QoS-minded utility proportional fairness framework for resource allocation for the cells of a cellular communication system that was introduced in Chap. 3. The distributed architecture was composed of a EURA optimization that allocated the UE rates by the eNB and an IURA optimization that assigned application rates by the UEs. Not only did we prove that the proposed distributed resource allocation architecture's EURA and IURA optimization problems are convex and solved them through the Lagrangian of their dual problem, but also we proved the optimality of the rate assignments. We showed that under abundance of resources, the resource allocation assigns more resources to the UEs with bad channel conditions, in order to meet their QoS requirements for their applications. This is in light of the fact that under a bad channel, lower modulation orders and coding schemes can be used, which reduces the spectrum efficiency. On the other hand, when the resources are constrained, more resources are allocated to UEs with good channel conditions so as to meet their bit rate requirements. Ultimately, we performed simulations in MATLAB to show the application of the proposed distributed resource allocation architecture to a cellular communication system.

References

1. J. Reed, N. Tripathi, *Cellular Communications: A Comprehensive and Practical Guide.* (Wiley-IEEE Press, 2014)
2. M. Richards, J. Scheer, W. Holm, *Principles of Modern Radar* (SciTech Publishing, 2010)
3. M. Ghorbanzadeh, A. Abdelhadi, C. Clacy, *Cellular Communications Systems in Congested Environments Resource Allocation and End-to-End Quality of Service Solutions with MATLAB* (Springer, Berlin, 2017)
4. M. Ghorbanzadeh, Resource allocation and end-to end quality of service for cellular communications systems in congested and contested environments. Ph.D. Thesis, Virginia Tech, 2015
5. J.G. Proakis, *Digital Communications* (McGraw Hill, New York, 2007)
6. M. Ghorbanzadeh, E. Visotsky, P. Moorut, W. Yang, C. Clancy, Radar inband and out-of-band interference into LTE macro and small cell uplinks in the 3.5 GHz band, in *2015 IEEE Wireless Communications and Networking Conference (WCNC)*, 2015
7. M. Ghorbanzadeh, E. Visotsky, P. Moorut, W. Yang, C. Clancy, Radar in-band interference effects on macrocell LTE uplink deployments in the U.S. 3.5 GHz band, in *2015 International Conference on Computing, Networking and Communications (ICNC)*, 2015
8. M. Ghorbanzadeh, E. Visotsky, P. Moorut, W. Yang, C. Clancy, Radar interference into LTE base stations in the 3.5 GHz band. Phys. Commun. **20**, 33–47 (2016)
9. S. Zarei, Channel coding and link adaptation. Seminar LTE: Der Mobilfunk der Zukunft, 2009
10. H. Shajaiah, M. Ghorbanzadeh, A. Abdelhadi, C. Clancy, Application-aware resource allocation based on channel information for cellular networks, in *2019 IEEE Wireless Communications and Networking Conference (WCNC)*, pp. 1–6, 2019
11. M. Ghorbanzadeh, A. Abdelhadi, C. Clancy, Application-aware resource allocation of hybrid traffic in cellular networks. IEEE Trans. Cogn. Commun. Netw. **3**(2), 226–241 (2017)
12. Y. Chen, M. Ghorbanzadeh, K. Ma, C. Clancy, R. McGwier, A hidden Markov model detection of malicious android applications at runtime, in *2014 23rd Wireless and Optical Communication Conference (WOCC)*, 2014
13. M. Ghorbanzadeh, A. Abdelhadi, C. Clancy, A utility proportional fairness radio resource block allocation in cellular networks, in *IEEE International Conference on Computing, Networking and Communications (ICNC)*, 2015
14. M. Ghorbanzadeh, Y. Chen, K. Ma, C. Clancy, R. McGwier, A neural network approach to category validation of android applications, in *IEEE Conference on Computing, Networking, and Communications (ICNC)*, 2013
15. M. Ghorbanzadeh, Y. Chen, C. Clancy, Fine-grained end-to-end network model via vector quantization and hidden Markov processes, in *IEEE Conference on Communications (ICC)*, 2013
16. M. Ghorbanzadeh, A. Abdelhadi, C. Clancy, A utility proportional fairness bandwidth allocation in radar-coexistent cellular networks, in *Military Communications Conference (MILCOM)*, 2014

17. G. T. . V9.0.0, Further advancements for E-UTRA physical layer aspects. Measuring of Heterogeneous Wireless and Wired Networks, 2012
18. Frame structure—downlink (2009). http://www.sharetechnote.com/
19. A. Ghosh, R. Ratasuk, Essentials of LTE and LTE-A. The Cambridge Wireless Essentials Series (2011)
20. S. Boyd, L. Vandenberghe, *Introduction to Convex Optimization with Engineering Applications* (Cambridge University Press, Cambridge, 2004)
21. F. Kelly, A. Maulloo, D. Tan, Rate control in communication networks: shadow prices, proportional fairness and stability. J. Oper. Res. Soc. **49**(3), 237–252 (1998)
22. S. Low, D. Lapsley, Optimization flow control, I: basic algorithm and convergence. IEEE/ACM Trans. Netw. **7**(6), 861–874 (1999)

Chapter 6
Propagation Modeling

6.1 Introduction

Chapter 5 introduced the channel modeling to the resource allocation presented in Chap. 3. The channel modeling, as shown in Chap. 5, relies on the concept of Signal-to-Noise Ratio (SNR).

While Chap. 5 leveraged some typical numbers for its simulations presented in Sect. 5.7 of that chapter, to perform a resource allocation in a wireless network in the field, as depicted in the next chapter, leveraging a realistic propagation model to represent the radio environment map (REM) is essential.

The propagation modeling translates directly to SNR, depicted in Sect. 5.5 of Chap. 5, used in the formulation of the resource allocation optimization in Chap. 5. To make the propagation model conducive to a real work wireless network, taking the effect of terrain elevation is of high consequence. Simple propagation models such as the free space pathloss (FSPL) [1–3], as shown in Eq. (6.1) where $X = -87.55, -27.55, 32.44, 92.45$ for (f, d) in (kHz, m), (MHz, m), (MHz, km), and (GHz, km), or Egli propagation model [4] amongst others do not consider a specific propagation path. For such models, should the reception point be over a water span equidistant from a TX with respect to that of another reception point in an urban environment, they both produce equal pathloss which is intuitively wrong. To provide a fine-grained REM, pathloss and clutter loss are great combinations; nonetheless, in this chapter, we focus on application of a terrain dependent pathloss model known as the Irregular Terrain Model (ITM) [5–9].

$$L_{p,dB} = 20(log_{10}(f) + log_{10}(d)) + X \qquad (6.1)$$

ITM was originally developed by Longley and Rice as a model intended for use in a large number of scenarios such as antenna heights and frequency. It can be used in an area prediction mode (APM) which does not rely on specific TX-RX terrain but on an overall terrain change in a geographic area. Moreover, it may be leveraged

© The Author(s), under exclusive license to Springer Nature Switzerland AG 2022
M. Ghorbanzadeh, A. Abdelhadi, *Practical Channel-Aware Resource Allocation*,
https://doi.org/10.1007/978-3-030-73632-3_6

in a point-to-point mode (P2PM) which relies on leveraging a terrain database (DB) to extract the precise elevation values on the path between the TX and the RX. In this chapter, we focus on using the P2PM since it uses the precise elevation on a path between the TX and the RX which is our desire as mentioned in the previous paragraph.

Hereinafter, a reference to ITM implies the P2PM ITM and not the APM ITM. At a high level, ITM requires the elevations of equidistant points on the great circle path between the TX and the RX as well as the TX and the RX Above Ground Level (AGL) antenna heights to return the pathloss.

The calculation of the geographic coordinates along the great circle path relies on various geodesic algorithms which are described in Sect. 6.2. Moreover, the elevations of the aforesaid geographic coordinates rely on using elevation DBs and elevation extraction algorithms, respectively, discussed in Sects. 6.3 and 6.3.1. Besides, while one can leverage typical climate code and refractivity values [10], they can be extracted from the DBs with relevant extraction algorithms presented in Sect. 6.3.2. Section 6.4 includes the computation of the ITM pathloss, Sect. 6.5 includes setting up of Python code to run the ITM pathloss, and Sect. 6.6 summarizes this chapter.

6.2 Geodesic Calculations

This section presents the geodesic computation algorithms which help contribute to ITM input parameters, i.e. the calculation of the geographic coordinates on the great circle path between the TX and the RX. The computations include Forward Vincenty [11], used to calculate the geographic coordinates at a given distance and azimuth from another geographic coordinates, and Inverse Vincenty [11], utilized to compute the distance d, forward azimuth α_F, and backward azimuth α_B between two geographic $\mathbf{S_1} = (\phi_1, \xi_1)$ and $\mathbf{S_2} = (\phi_2, \xi_2)$, where the first/second element of the tuple is latitude/longitude in degrees (deg) on the Earth. The Inverse Vincenty is done according to Algorithm 15, while a geographic coordinate $\mathbf{S_k} = (\phi_k, \xi_k)$ at distance/azimuth d_k/α_{12} from $\mathbf{S_1}$ can be obtained using Algorithm 16. Together, these two algorithms can be used to compute the great circle path coordinates using Algorithm 17. An implementation of Algorithms 15 and 16 appears in [12].

Vincenty intended to express geodesic existing algorithms on an ellipsoid so as to minimize the program length. [11] originally used a Wang 720 desk calculator, equipped with a few kilobytes (kB) memory. To achieve a good accuracy, Algorithms 16 and 15 use the Legendre, Bessel, and Helmert classical solutions based upon the auxiliary sphere. Vincenty leveraged the formulation of Rainsford. Initially, Legendre had shown that an ellipsoidal geodesic could be mapped to the auxiliary sphere by mapping the latitude to a reduced one and equating the azimuth of the great circle to that of the geodesic. The ellipsoidal longitude and the distance are then derived in terms of the spherical longitude and the great circle arc

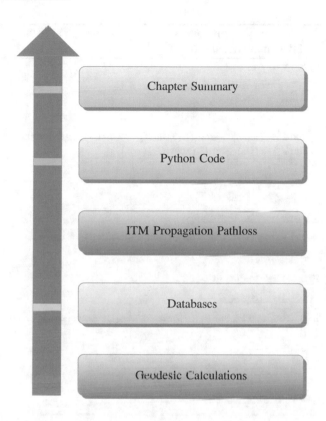

length via several integrations. Next, Bessel and Helmert provided convergent series corresponding to those integrals in order to do the computation with an arbitrary accuracy.

Vincenty, in an effort to minimize the program length, expanded the aforementioned series via parameterizing the first term of each series and truncated them to $O(f^3)$, which produced compact expressions of the distance and longitude integral operations. Next, the expressions were formulated as Horner form which enables polynomials evaluation via a single memory register. Ultimately, an iterative technique solved the equations in the forward and inverse methods as depicted in Algorithms 16 and 15, respectively.

Inverse Vincenty Algorithm 15 finds the distance d in km and forward/backward azimuth α_{12}/α_{21} in deg from geographic coordinates (ϕ_1, ξ_1) to geographic coordinates (ϕ_2, ξ_2). On the other hand, Forward Vincenty Algorithm 16 finds the geographic coordinates $\mathbf{S_k}$ deg at a distance d_k km and forward azimuth α_{12} deg from another geographic coordinates $\mathbf{S_1}$ deg. In these algorithms, $a = 6378.1370$ is the earth semi-major axis and $f = \frac{1}{298.257223563}$ is the earth ellipsoid flattening according to World Geological Survey (WGS) 1984 (WGS84).

Algorithm 15 Inverse Vincenty finds (ϕ_1, ξ_1) to (ϕ_2, ξ_2) distance d km and forward/backward azimuth α_{12}/α_{21} deg

---

Ensure: $a = 6378.1370$, $f = \frac{1}{298.257223563}$, $\lambda_0 = 9e^{40}$

1: $U_i = atan((1 - f)\tan(\phi_i \frac{\pi}{180})), i = 1, 2$

2: $\lambda = |(\xi_2 - \xi_1)\frac{\pi}{180}|$

3: **while** $|\lambda - \lambda_0| > e^{-12}$ **do**

4:    $\lambda_0 = \lambda$

5:    $\sin(\sigma) = [(\cos(U_2)\sin(\lambda))^2 + (\cos(U_1)\sin(U_2) - \cos(U_2)\sin(U_1)\cos(\lambda))^2]^{0.5}$

6:    $\cos(\sigma) = \sin(U_1)\sin(U_2) + \cos(U_1)\cos(U_2)\cos(\lambda)$

7:    $\sigma = atan2(\frac{\sin(\sigma)}{\cos(\sigma)})$

8:    $\sin(\alpha) = \frac{\cos(U_1)\cos(U_2)\sin(\lambda)}{\sin(\sigma)}$

9:    $\cos(2\sigma_m) = \cos(\sigma) - \frac{2\sin(U_1)\sin(U_2)}{1 - \sin^2(\alpha)}$

10:    $C = \frac{f\{4 + f[4 - 3(1 - \sin^2(\alpha))]\}(1 - \sin^2(\alpha))}{16}$

11:    $\lambda = |(\xi_2 - \xi_1)\frac{\pi}{180}| + f(1 - C)\sin(\alpha)\{\sigma + C\sin(\sigma)[\cos(2\sigma_m) + C\cos(\sigma)(-1 + 2\cos^2(2\sigma_m))]\}$

12: **end while**

13: $u^2 = \frac{(a^2 - b^2)(1 - \sin^2(\alpha))}{b^2}$

14: $A = 1 + \frac{u^2\{4096 + u^2[-768 + u^2(320 - 175u^2)]\}}{16384}$

15: $B = \frac{u^2\{256 + u^2[-128 + u^2(74 - 47u^2)]\}}{1024}$

16: $\Delta\sigma = B\sin(\sigma)\{\cos(2\sigma_m) + \frac{B}{4}[\cos(\sigma)(-1 + 2\cos^2(2\sigma_m)) - \frac{B}{6}\cos(2\sigma_m)(-3 + 4\sin^2(\sigma))(-3 + 4\cos^2(2\sigma_m))]\}$

17: $d = aA(1 - f)(\sigma - \Delta\sigma)$

18: $\alpha_{12} = \frac{[(atan2(\frac{\cos(U_2)\sin(\lambda)}{\cos(U_1)\sin(U_2) - \sin(U_1)\cos(U_2)\cos(\lambda)}) + 2\pi) \mod 2\pi]180}{\pi}$

19: $x = atan2(\frac{\cos(U_1)\sin(\lambda)}{-\cos(U_2)\sin(U_1) + \sin(U_2)\cos(U_1)\cos(\lambda)})$

20: $\alpha_{21} = \frac{[(x + 2\pi U(\pi - x) - \pi U(x - \pi) + 2\pi) \mod 2\pi]180}{\pi}$

This is shown in Fig. 6.1 where the dashed line represents a spherical model and the earth ellipsoid with axes are shown. Since the ITM propagation relies on the elevations of the great circle path between a TX and a RX [10, 13–16], in order to obtain the elevations, computation of the geographic coordinates along the great circle path is required. This can be accomplished by means of Algorithm 17 which finds the great circle coordinates from the geographic coordinates S_1 to the geographic coordinates S_2 by first finding the distance d in km and azimuth θ in deg from the geographic coordinates S_1 to S_2 using Algorithm 15. Next, the calculated distance d is leveraged to compute the number of points n on the great circle path with a spacing of Δd_0 km and a maximum of n_{Max}.

Then, for the n great circle points, the inter-distance $\Delta d = \frac{d}{n-1}$ in km is computed. Finally, Algorithm 16 is leveraged employed to calculate the n points S_{ik}, in which subscript i refers to "intermediate" to imply intermediate points between S_1 to S_2, at distances $k\Delta d$ where $k \in \{0, 1\ldots, n\}$. An implementation of Algorithm 17 appears in [12]. Now that the geographic coordinates on the great circle path are computed, their coordinates S_{ik} can be leveraged to extract the elevation using an elevation DB and relevant algorithms as Sect. 6.3.1. Moreover, ITM can be used with typical climate and refractivity index values [10] or they

Algorithm 16 Forward Vincenty gets point $\mathbf{S_k}$ deg at distance d_k km and forward azimuth α_{12} deg from point $\mathbf{S_1}$ deg

Ensure: $a = 6378.1370$, $f = \frac{1}{298.257223563}$, $\sigma_0 = e^{40}$

1: $U_1 = atan((1-f)\tan(\frac{\phi_1 \pi}{180}))$

2: $\sigma_1 = atan2(\frac{tan(U_1)}{cos(\alpha_1)})$

3: $\sin(\alpha) = \cos(U_1)\sin(\frac{\alpha_{12}\pi}{180})$

4: $u^2 = \frac{a^2(1-\sin^2(\alpha))[1-(1-f)^2]}{((1-f)a)^2}$

5: $A = 1 + \frac{u^2\{4096+u^2[-768+u^2(320-175u^2)]\}}{1024}$

6: $B = \frac{u^2\{256+u^2[-128+u^2(74-47u^2)]\})}{1024}$

7: **if** $d_k > e^{-20}$ **then**

8: $\quad \sigma = \frac{d_k}{(1-f)aA}$

9: $\quad$ **while** $|\sigma - \sigma_0| > e^{-12}$ **do**

10: $\quad\quad \sigma_0 = \sigma$

11: $\quad\quad 2\sigma_m = 2\sigma_1 + \sigma$

12: $\quad\quad \Delta\sigma = B\sin(\sigma)\{\cos(2\sigma_m) + 0.25B[\cos(\sigma)(-1+2cos^2(2\sigma_m) - \frac{B}{6}\cos(2\sigma_m)(-3+4\sin^2(\sigma))(-3+4\cos^2(2\sigma_m))]\}$

13: $\quad\quad \sigma = \frac{d_k}{(1-f)aA} + \Delta\sigma$

14: $\quad$ **end while**

15: $\quad y = (1-f)[\sin^2(\alpha) + (\cos(U_1)\cos(\sigma)\cos(\frac{\alpha_{12}\pi}{180}) - \sin(U1)\sin(\sigma))^2]^{0.5}$

16: $\quad \phi_k = atan2(\frac{\sin(U_1)\cos(\sigma)+\cos(U_1)\sin(\sigma)\cos(\frac{\alpha_{12}\pi}{180})}{y})\frac{180}{\pi}$

17: $\quad \lambda = atan2(\frac{\sin(\sigma)\sin(\frac{\alpha_{12}\pi}{180})}{\cos(U_1)\cos(\sigma)-\sin(U_1)\sin(\sigma)\cos(\frac{\alpha_{12}\pi}{180})})$

18: $\quad C = \frac{f(1-\sin^2(\alpha))\{4+f[4-3(1-\sin^2(\alpha))]\}}{16}$

19: $\quad L = \lambda - (1-C)f\sin(\alpha)\{\sigma + C\sin(\sigma)[\cos(2\sigma_m) + C\cos(\sigma)(2\cos^2(2\sigma_m)-1)]\}$

20: $\quad \xi_k = \frac{180L+\xi_1\pi}{\pi}$

21: **end if**

Algorithm 17 $\mathbf{S_1}$ - $\mathbf{S_2}$ great circle points $S_{i,k}$ calculation where i stands for "intermediate"

Ensure: $k = 0$, set Δd_0 and n_{Max}.

1: Find $\mathbf{S_1}$ - $\mathbf{S_2}$ distance d, azimuth θ (Algorithm 15).

2: $n = max\{min\{\frac{d}{\Delta d_0}, n_{Max}\}, 1\}$.

3: $\Delta d = \frac{d}{n-1}$

4: **for** $k \in \{0, 1, \ldots, n\}$ **do**

5: $\quad$ Find $\mathbf{S_{ik}} = (\phi_{ik}, \xi_{ik})$, $k = 0, 1, \ldots, n-1$, at distance $k\Delta d$ km and azimuth (Algorithm 16).

6: **end for**

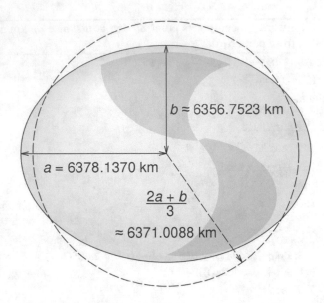

Fig. 6.1 The earth ellipsoid in WGS84

can be extracted from DBs as shown in Sect. 6.3. The implementation in [12] sets
$n_{Max} = 1501$ points and $\Delta d_0 = 0.03$ km, while selection of any other positive
spacing and point quantity is valid.

As mentioned earlier, an implementation of these geodesic methods can be
found in the open source git repository [12] with the instructions to obtain the
code base in Sect. 6.5. One you obtain the code, you may see the implementation
of Algorithm 15 at the *src/harness/* `reference_models`*/geo/vincenty.py* whose
function *GeodesicDistanceBearing* annotated in Fig. 6.2 implements Algorithm 15
to return the distance, forward azimuth, and backward azimuth from one geographic
coordinate to another, the inputs to the function.

On the other hand, the function *GeodesicPoint* at *src/harness/* `reference`
`_models`*/geo/vincenty.py* implements Algorithm 16, i.e. the forward Vincenty, to
return the geographic coordinates at an input distance in km and forward azimuth
in deg from another input geographic coordinates as shown in Fig. 6.3. While
this function can accept a vector of input distances to return a vector of output
geographic coordinates, there is a scalar version of the aforesaid function which
inputs a single distance in km and returns a pair of geographic coordinates. This
function is called *GeodesicPoint* and is equivalent to *GeodesicPoints* when there is
only one element in the input distance vector.

Moreover, Algorithm 17 implementation is presented in function *GeodesicSampling* of *src/harness/* `reference_models`*/geo/vincenty.py*, depicted in Fig. 6.4,
which relies on the previously mentioned forward Vincenty implementation, i.e.
Algorithm 16, in function *GeodesicPoints*.

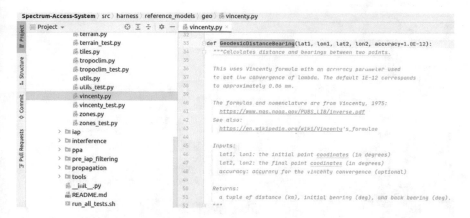

Fig. 6.2 Inverse Vincenty, Algorithm 15, is implemented in function *GeodesicDistanceBearing* at *src/harness/reference_models/geo/vincenty.py* of the open source git repository [12]

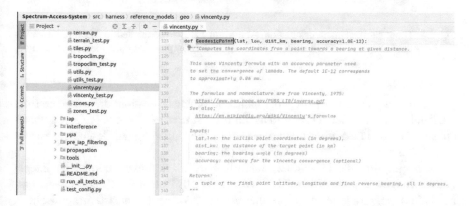

Fig. 6.3 Forward Vincenty, Algorithm 16, is implemented in function *GeodesicPoints* at *src/harness/reference_models/geo/vincenty.py* of the open source git repository [12]

6.3 Databases

This section presents the DBs leveraged in the computation of the ITM propagation loss. These are terrain elevation, surface refractivity index, and climate as shown in Sect. 6.3.1 and 6.3.2.

6.3.1 Terrain Profile

Hereinafter, the elevation profile refers to the Above Mean Sea Level (AMSL) elevation of a great circle path $\{S_{ik}, k = 0, 1, \ldots, n_{Max}\}$, explained in Sect. 6.2.

```
Spectrum-Access-System ) src ) harness ) reference_models ) geo ) vincenty.py
  Project ▾              ⊙ Σ ÷ | ✿ — | vincenty.py
            terrain.py              300   L = (lambda - (1. - C) * f * sinalpha
            terrain_test.py         301       * (sigma + C * sin_sigma
            tiles.py                302       * (cos_twosigma + C * cos_sigma
            tropoclim.py            303       * (-1. + 2. * cos_twosigma**2))))
            tropoclim_test.py       304   L2 = L + L1
            utils.py                305
            utils_test.py           306   num = sinalpha
            vincenty.py             307   den = -sin(U1) * sin_sigma + cos(U1) * cos_sigma * cos(alpha1)
            vincenty_test.py        308   alpha2 = np.arctan2(num, den)
            zones.py                309   alpha2 = (alpha2 + 3.*pi) % (2.*pi)
            zones_test.py           310
        > ▢ iap                     311   if isinstance(distances_km, np.ndarray):
        > ▢ interference            312       return np.degrees(phi2), np.degrees(L2), np.degrees(alpha2)
        > ▢ ppa                     313   else:
        > ▢ pre_iap_filtering       314       return list(np.degrees(phi2)), list(np.degrees(L2)), list(np.degrees(alpha2))
        > ▢ propagation             315
        > ▢ tools                   316
            __init__.py             317   def GeodesicSampling(lat1, lon1, lat2, lon2, num_points):
            README.md               318       """Returns a geodesic between 2 points defined as equally spaced points.
            run_all_tests.sh        319
            test_config.py          320       Inputs:
        > ▢ testcases               321           lat1, lon1 : the initial point coordinates.
            __init__.py             322           lat2, lon2 : the final point coordinates.
            CHANGELOG.md            323           num_points : number of points to use (must be >=2)
            common_strings.py       324
            common_types.py         325       Returns:
            CRLServer-README.md     326           A tuple (lats, longs) of ndarray vector defining points along the
            database.py             327           geodesic. The two input locations are first and last points.
                                    328       """
```

Fig. 6.4 The great circle path calculation, Algorithm 17, is implemented in function *Geodesic-Sampling* at *src/harness/* `reference_models` */geo/vincenty.py* of the open source git repository [12]

The aforesaid great circle path geographic coordinates elevations are extracted from a Digital Elevation Model (DEM) DB, an open source of which is purveyed with 1 arc second (arcs) resolution 1 deg tiles [17] by the United Stated Geological Survey (USGS), aka the National Elevation Data (NED).

Each elevation tile is uniquely identified by its northwest corner, an integer latitude/longitude preceded by "N"/"W" or "S"/"E" when the tile northwest corner latitude/longitude is in the northern/western or southern/eastern hemisphere, respectively. As a case in point, tiles N37W078 and S04E171 cover, respectively, the latitude/longitude range $(36, 37)/[-78, -77]$ and $(-5, -4)/[171, 172]$ deg. For the kth point $\mathbf{S_{ik}} = (\phi_{i,k}, \xi_{i,k})$ along the great circle path, the tile of interest is $(\lfloor \xi_{i,k} \rfloor, \lceil \phi_{i,k} \rceil)$. Each 1 deg tile includes 3612 rows and 3612 columns, where rows/columns $6 - 3600$ uniquely form 1 deg worth of elevation data spaced at $\frac{1}{3600}$ asec.

Then, the point $\mathbf{S_{i,k}}$ elevation $e_{i,k}$ from the tile specified by $\lceil \phi_{i,k} \rceil$ and $\lfloor \xi_{i,k} \rfloor$, denoted as $E_{\lceil \phi_{i,k} \rceil, \lfloor \xi_{i,k} \rfloor}$ can be obtained by bilinear interpolation [18] of the point's 4 neighbors, namely upleft, upright, downleft, and downright neighbor elevation $E_{\lceil \phi_{i,k} \rceil, \lfloor \xi_{i,k} \rfloor}[n]$, $E_{\lceil \phi_{i,k} \rceil, \lfloor \xi_{i,k} \rfloor}[n+3612]$, $E_{\lceil \phi_{i,k} \rceil, \lfloor \xi_{i,k} \rfloor}[n+1]$, and $E_{\lceil \phi_{i,k} \rceil, \lfloor \xi_{i,k} \rfloor}[n+3613]$ according to Eq. (6.2), an implementation of which appears in [12].

$$e_{i,k} = E_{\lceil \phi_{i,k} \rceil, \lfloor \xi_{i,k} \rfloor}[n](1 - \Delta Y)(1 - \Delta X) + E_{\lceil \phi_{i,k} \rceil, \lfloor \xi_{i,k} \rfloor}[n + 1]\Delta X(1 - \Delta Y)$$

$$+ E_{\lceil \phi_{i,k} \rceil, \lfloor \xi_{i,k} \rfloor}[n + 3612]\Delta Y(1 - \Delta X)$$

```
Spectrum-Access-System  src  harness  reference_models  geo  terrain.py
Project ▾                        ⊘ Σ ÷ ✿ —    terrain.py
        refractivity.py                166    def GetTerrainElevation(self, lat, lon, do_interp=True):
        refractivity_test.py           167    ●   """Retrieves the elevation for one or several points.
        terrain.py                     168
        terrain_test.py                169        This function is vectorized for efficiency.
        tiles.py                       170
        tropoclim.py                   171        Inputs:
        tropoclim_test.py              172            lat, lon (scalar or iterables such as list or ndarray): coordinates of
        utils.py                       173               points to read (degrees).
        utils_test.py                  174            do_interp (bool): if True, the elevation is bilinearly interpolated.
        vincenty.py                    175
        vincenty_test.py               176        Returns:
        zones.py                       177            the terrain elevation(s) as:
        zones_test.py                  178              - a scalar if the input point is scalar.
      > ▭ iap                          179              - a ndarray of elevations, if the input lat/lon are iterables.
                                       180        """
```

Fig. 6.5 Point elevation calculation is implemented in function *GeodesicSampling* at *src/harness/*`reference_models`*/geo/terrain.py* of the open source git repository [12]

$$+ E_{\widetilde{\lceil \phi_{i,k} \rceil}, \lfloor \xi_{i,k} \rfloor}[n + 3613] \Delta Y \Delta X \tag{6.2}$$

The number inside [] is the index of the element from the tile $E_{\widetilde{\lceil \phi_{i,k} \rceil}, \lfloor \xi_{i,k} \rfloor}$ starting at 0 as each 3612×3612 tile is a one dimensional vector version of the two dimensional matrix, arranged row-wise. For instance, the second row and third column (indexes 1 and 2) can be located at the one dimensional equivalent $(2-1) \times 3612 + 1$. The interpolation weights are $R = 3600(\lceil \phi_{i,k} \rceil - \phi_{i,k}) + 5.5, C = 3600(\xi_{i,k} - \lfloor \xi_{i,k} \rfloor) + 5.5, \Delta Y = R - \lfloor R \rfloor, \Delta X = C - \lfloor C \rfloor$, and $n = \lfloor C \rfloor + 3612 \lfloor R \rfloor$. The implementation of elevation computation appears in function *GetTerrainElevation* of *src/harness/*`reference_models`*/geo/terrain.py*, depicted in Fig. 6.5 which returns elevation of an input geographic coordinates.

Besides, the implementation of elevation of the great circle path between two geographic coordinates appears in function *TerrainProfile* of *src/harness/*`reference_models`*/geo/terrain.py*, depicted in Fig. 6.6, which computes the great circle according to Algorithm 17 and calculates/returns the elevations at the great circle geographic coordinates by relying on the formerly mentioned function *GetTerrainElevation*. Next, Sect. 6.3.2 describes the parameters climate and surface refractivity index to be used by the ITM propagation model.

6.3.2 Surface Refractivity and Climate

Climate DB, 360×720 matrix $\widetilde{A}$ appearing in the International Telecommunications Union Radiocommunication sector (ITU-R) Radio wave propagation (ITU-R) recommendation 2001 (ITU-R P.2001) [19] contains global climate codes, where the values are from the set $\{0, 1, 2, 3, 4, 5, 6, 7\}$ whose elements correspond, respectively, to the elements of the set {missing data, equatorial climate, subtropical, maritime tropical, desert, continental temperate, land maritime temperate, sea maritime temperate}, at a rectangular grid points with latitudes $89.75 - 0.5n, n =$

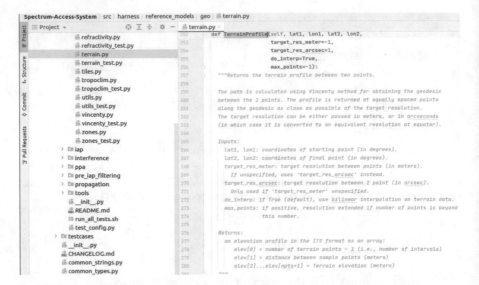

Fig. 6.6 Great circle elevation calculation is implemented in function *TerrainProfile* at *src/harness/`reference_models`/geo/terrain.py* of the open source git repository [12]

$0, 1, \ldots, 359$ and longitudes $-179.75 + 0.5m, m = 0, 1, \ldots, 719$. On the other hand, the refractivity index appears in ITU-R P.452 [20], indicating how much waves propagate per latitude on the globe, is a 121×241 matrix $\widetilde{B}$ with latitudes $90 - 1.5n, n = 0, 1, \ldots, 120$ and longitudes $(1.5m + 180 \mod 360) - 180, m = 0, 1, \ldots, 240$. Climate parameter K_{Climate} of the great circle path from the point (ϕ_1, ξ_1) to (ϕ_2, ξ_2) deg, at distance d and azimuth α_{12} is computed by the following steps, succinctly summarized in Algorithm 18, according to [12].

1. Calculating midpoint coordinates (ϕ_M, ξ_M) at distance $0.5d$ km and azimuth α_{12} from (ϕ_1, ξ_1) @! Algorithm 16.
2. Finding midpoint $\beta_{\phi_M \xi_M}$ climate code from the aforesaid climate database, at row ($\lfloor \frac{89.75 - \phi_M}{0.5} + 0.5 \rfloor$) and column ($\lfloor \frac{\xi_M + 179.75}{0.5} + 0.5 \rfloor$)) closest to the midpoint coordinates.
3. A $\beta_{\phi_M \xi_M}$ which is extracted as 0 is mapped to 7 for computation purposes and a $\beta_{\phi_M \xi_M}$ of 7 (whether being sea maritime temperate or a 7-mapped missing data category) requires a further step to compute the same parameter $\beta_{\phi_i \xi_i}, i = 1, 2$ for the points (ϕ_1, ξ_1) and (ϕ_2, ξ_2) and choosing the minimum of these end-point $\beta_{\phi_i \xi_i}$ values as k_{Climate}; otherwise, the midpoint $K_{\text{Climate}} = \beta_{\phi_M \xi_M}$.
4. Use Eq. (6.3) to interpolate from neighbors, aka bilinear interpolation [18].

$$k_{\text{Refractivity}} = \widetilde{B} \left(\frac{90 - \phi_M}{1.5} + 1, \frac{\xi_M}{1.5} + 1 \right) \left(\frac{90 - \phi_M}{1.5} ! \lfloor \frac{90 - \phi_M}{1.5} \rfloor \right) \left(\frac{\xi_M}{1.5} - \lfloor \frac{\xi_M}{1.5} \rfloor \right)$$

$$+\widetilde{B}\left(\frac{90-\phi_M}{1.5},\frac{\xi_M}{1.5}\right)\left(1-\frac{90-\phi_M}{1.5}+\lfloor\frac{90-\phi_M}{1.5}\rfloor\right)\left(1-\frac{\xi_M}{1.5}+\lfloor\frac{\xi_M}{1.5}\rfloor\right)$$

$$+\widetilde{B}\left(\frac{90-\phi_M}{1.5},\frac{\xi_M}{1.5}+1\right)\left(1-\frac{90-\phi_M}{1.5}+\lfloor\frac{90-\phi_M}{1.5}\rfloor\right)\left(\frac{\xi_M}{1.5}-\lfloor\frac{\xi_M}{1.5}\rfloor\right)$$

$$+\widetilde{B}\left(\frac{90-\phi_M}{1.5}+1,\frac{\xi_M}{1.5}\right)\left(\frac{90-\phi_M}{1.5},\frac{\xi_M}{1.5}\right)\left(1-\frac{\xi_M}{1.5}+\lfloor\frac{\xi_M}{1.5}\rfloor\right)$$

Algorithm 18 Surface refractivity and climate computation

1: Find S_1 - S_2 distance d, azimuth θ (Algorithm 15).
2: Find (ϕ_M, ξ_M) at distance, azimuth $0.5d$, θ from S_1 (Algorithm 16).
3: **for** $i \in \{1, 2, M\}$ **do**
4:     $\beta_{\phi_i\xi_i} = \widetilde{A}(\lfloor\frac{89.75-\phi_i}{0.5}+0.5\rfloor, \lfloor\frac{\xi_i+179.75}{0.5}+0.5\rfloor)$
5: **end for**
6: Use equation (6.3) to get K_{Climate}
7: **if** $\xi_M < 0$ **then**
8:     $\xi_M = \xi_M + 360$
9: **end if**
10: Apply equation (6.3) to obtain $k_{\text{Refractivity}}$.

The DBs representing the climate code from the ITU-R P.2001 and the refractivity index from ITU-R P.452 are shown in Fig. 6.7a,b respectively. These are the raw DB values at latitude/longitude values $89.75 - 0.5n$, $n = 0, 1, \ldots, 359$ and longitudes $-179.75+0.5m$, $m = 0, 1, \ldots, 719$ for climate code and $90-1.5n$, $n = 0, 1, \ldots, 120$ and longitudes $(1.5m + 180 \mod 360) - 180$, $m = 0, 1, \ldots, 240$ for surface refractivity.

$$k_{\text{Climate}} = \min\left\{\overbrace{\beta_{\phi_i\xi_i}\left(U(7-\beta_{\phi_i\xi_i})-U(\beta_{\phi_i\xi_i}-1)\right)+7\delta(\beta_{\phi_i\xi_i})}^{\gamma_{\phi_i\xi_i},\,i\in\{1,2,M\}}\right\}$$

$$\times\prod_{j=1}^{2}\delta(\gamma_{\phi_j\xi_j}-7)+\gamma_{\phi_m\xi_m}U(6-\gamma_{\phi_m\xi_m}) \tag{6.3}$$

The bilinear interpolation [18] shown in Eq. (6.2) is depicted in Fig. 6.8 which shows the point for which the data computation is needed, whether it is a geographic coordinate for which elevation is desired or one whose refractivity is wanted, as the black circle whose immediate northwest northeast, southwest, and southeast neighbors are points shown in yellow, purple, green, and pink, geographic coordinates closest to those of the black point in the middle of Fig. 6.8.

The aforesaid neighbors have values are the closest points, to the black point for which a computation (for elevation or refractivity) is desired, that have values in the underlying DB such as the USGS 1 asec elevation [17] as explained in Sect. 6.3.1 or

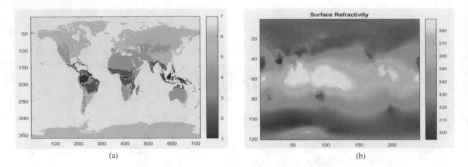

(a)   (b)

Fig. 6.7 Surface refractivity, and climate databases. (**a**) Climate Parameter for $\phi_n = 89.75 - 0.5n, n = 0, 1, \ldots, 359$ and $\xi_n = 0.5m - 179.75, m = 0, 1, \ldots, 719$. (**b**) Surface Refractivity

Fig. 6.8 The bilinear interpolation to obtain elevation or refractivity

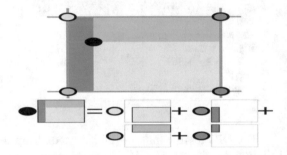

ITU-R P.452 [20] as expounded in Sect. 6.3.2. Next, the data value for each of the neighbors mentioned above are extracted from the underlying DB; these would be the AMSL elevations for the yellow, purple, green, and pink points, i.e. geographic coordinates, or their refractivity index. Next, Sect. 6.4 explains ITM propagation pathloss [21].

Afterwards, the extracted values for the neighbors are weighted by the areas of the rectangles diagonally opposite from the point to be assigned the weight. Such areas are identically colored to those of the neighbor points. For instance, the weight of the pink geographic coordinates, the southeast neighbor of the black point, is the area of the pink rectangle on the upper west corner of Fig. 6.8. As a case in point, for elevation extraction, the elevation of these points is multiplied by the specified areas. Finally, the products of the weighted neighbor values are added to stem out the ultimate result for the black point. In case of elevation or surface refractivity, those parameters are multiplied by their corresponding areas and the products added yield in the final elevation or surface refractivity. An implementation of refractivity of the great circle path between two geographic coordinates appears in function *Refractivity* of *src/harness/reference_models/geo/refractivity.py* which calculates/returns the refractivity at a geographic coordinates. Similarly, an implementation of the climate code calculation appears in the function *TropoClim* of *src/harness/reference_models/geo/tropoclim.py* as shown in Fig. 6.9. Next,

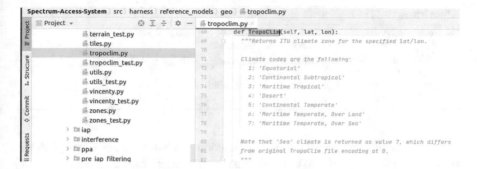

Fig. 6.9 Climate/Refractivity calculation is implemented in function *TropoClim/Refractivity* at *src/harness/reference_models /geo/tropoclim.py/src/harness/reference_models /geo/refractivity.py* of the open source git repository [12]

Sect. 6.3.2 describes the parameters climate and surface refractivity index to be used by the ITM propagation model.

6.4   ITM Propagation Pathloss

ITM [21], a statistical model based on the electromagnetic theory and derived from extensive sets of measurements, estimates radio propagation pathloss via diffraction and troposcatter phenomena. ITM provides accurate modeling for larger distances (order of km) as well as for rural areas where terrain AMSL elevation as opposed to man-made clutter constitutes the bulk of the environment. As mentioned earlier, ITM has two modes of operation, denoted as Point-to-Point Prediction mode (P2PM) and Area Prediction mode (APM) [21]. Hereinafter, usage of ITM refers to P2PM ITM. ITM has been adopted by some applications in the context of Citizen Radio Broadband Services (CBRS) [12] and has enjoyed availability of reference open-source code from both CBRS-related Winnforum [12] and National Telecommunications and Information Administration (NTIA) [22]. Besides, ITM as well as the nuances of its preparatory modules/parameters, mentioned in Sects. 6.3.1, 6.2, and 6.3.2, have been extensively vetted in by consistent and broad feedback from NTIA, Department of Defense (DOD), and various commercial entities of interest and all included end-to-end in the reference code provided by CBRS [12] which makes all the modules readily available. ITM relies on the AMSL elevations of the great circle path [21] succinctly referred to as the path profile (Sect. 6.3.1). To use ITM, one would extract the terrain profile from a TX to its RX along their great circle path, provide TX and RX antenna AGL heights, and compute the great circle path surface refractivity Sect. 6.3.2 and climate (Sect. 6.3.2), then ITM estimates the signal attenuation in decibels (dB).

ITM is applicable to the 20 MHz–20 GHz frequency band, the distance range 0.1–2000 km, antenna AGL height 0.5–3000 m, and a TX antenna with vertical

Algorithm 19 ITM high level algorithm

1: Compute TX-RX distance d and forward azimuth α_{12} using Algorithm 15.
2: Find great circle length n and points via Algorithm 17.
3: Find the great circle elevation profile $e_{i1}, \ldots, e_{in}$ using the USGS DB (Section 6.3.1) and bilinear interpolation in Equation (6.2).
4: Compute the refractivity and climate via Algorithm 18.
5: Compute ITM pathloss L_{ITM}.

or horizontal polarization [21]. ITM is a function of TX-RX distance as well as the signal variability in time. ITM usage preparation consumes various geodesic computations to account for Earth curvature; as such, any geographical calculations should account for nuances of Earth curvature. The algorithms for such computations were explained in Sect. 6.2. Moreover, ITM computation needs leveraging terrain elevation DBs, explained in Sect. 6.3.1, and refractivity/climate parameters 6.3.2. This is summarized in Algorithm 19. While ITM explicitly evaluates signal loss due to terrain AMSL elevation, losses due to vegetation, man-made structures, foliage, and ducting are present in the model to the extent that they have been present in the original measurements [21, 23]. The model contains some variability, expressed as confidence and reliability [21], intended to tacitly reflect ground clutter presence inasmuch as the variability parameters are based on measurements over paths which may subsume vegetation and man-made objects. Another limitation of the model is dependency of its predictions on digital terrain database and profile extraction methodology. For example, for a communications link, ITM adoption of a 30 asec database with 500 m spaced extracted elevation points can produce significantly different results as opposed to a 50 m inter-spaced extracted elevation points from a 3 asec database. Another limitation of the model is being devoid of precipitation-induced losses, time of the day, and season of the year [21]. An implementation of the ITM appears in function *CalcItmPropagationLoss* of *src/harness/reference_models/propagation/wf_itm.py* which calculates/returns the ITM pathloss from a TX at an input geographic coordinates and antenna AGL height and a RX at an input geographic coordinates and antenna AGL height as shown in Fig. 6.10. This function is the end-to-end function, which includes the great circle computation in Algorithm 17, elevation calculation in Sect. 6.3.1, and refractivity/climate calculation in Algorithm 18, and should be called via wrapper is presented in Sect. 6.5.2.

6.5   Python Code

The Python code version 2.7 [24], for Ubuntu flavor of Linux [25], is located at [12]. This repository includes code and DBs for evaluating the compliance of the Spectrum Access System (SAS) software, defined by the Federal Communications Commission (FCC) proceeding 12–354 as a system which authorizes access to the

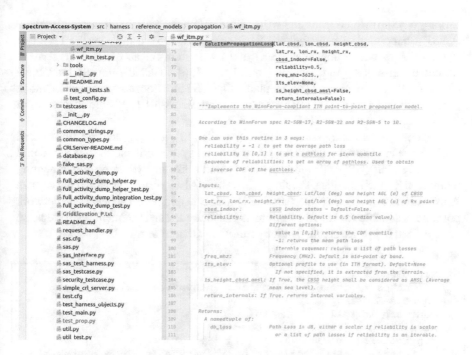

Fig. 6.10 ITM pathloss calculation is implemented in function *CalcItmPropagationLoss* at *src/harness/reference_models/propagation/wf_itm.py* of the open source git repository [12]

3550–3700 MHz CBRS. The code base includes geodesic calculations in Sect. 6.2, elevation calculation in Sect. 6.3.1, climate/refractivity computation in Sect. 6.3.2, and propagation computation in Sect. 6.4. The installation is provided in Sect. 6.5.1.

6.5.1   Code Base Installation

The code can be obtained as shown below. Next "pip" Python package management system [26] should be installed. Next, "libgeos" binary library [27] is needed to support Python shapely libraries. Then, [28] binary library supporting the Python XML libraries should be installed. One may need to configure a platform-specific C++ interpreter so as to complete the installation. Next, libgdal [29] binary library which handles geospatial data methods should be installed. Moreover, numpy [30] is leveraged by the project code and provides help with a variety of numerical methods. Moreover, [31] installation is required to handle the geographic shapes used in the program.

Moreover, pyJWT [32], pykml [33], cryptography and [34] binary libraries for handling KML geospatial files in Python are required. Also, Python Shape-file

Library (PyShp) which reads and writes Environmental Systems Research Institute (ESRI) Shape-files in pure Python. Besides, ftputil [35], a high level FTP client Python library which implements a virtual file system for accessing FTP servers, should be installed. Additionally, jsonschema [36] to handle JavaScript Object Notation (JSON), pyopenssl [37] to handle Open Secure Sockets Layer (SSL), mock [38] to handle/create mock software tests, and functools [39] library to support advanced functools are needed.

Besides, process and system utilities (psutil) [40], a cross-platform library that can retrieve information of running processes and system utilization such as Central Processing Unit (CPU), memory, network, and disks in Python by implementing many UNIX functionalities, should get installed. Also, portpicker which purveys an Application Programming Interface (API) to locate and return an available network port and is suitable for use from unit tests is required. The aforementioned steps are summarized below assuming that they run in an Ubuntu terminal.

```
1  git clone ...
        https://github.com/Wireless-Innovation-Forum/Spectrum
2  -Access-System.git
3  sudo apt install python-pip
4  sudo apt-get install libgeos-dev
5  pip install --user ...
        https://github.com/matplotlib/basemap/archive/master.zip
6  pip install GDAL
7  pip install numpy
8  pip install Shapely
9  pip install PyJWT
10 pip install pykml1
11 pip install cryptography
12 pip install pyshp
13 pip install ftputil
14 pip install jsonschema
15 pip install pyOpenSSL
16 pip install mock
17 pip install functools32
18 pip install psutil
19 pip install portpicker
```

Next, the USGS 1 asec elevation data and climate/refractivity data as explained in Sect. 6.3 are to be acquired. The repository is located at [41]. The full dataset is approximately 52 Gigabytes (GB). Moreover, retrieving the raw files from Large File Storage (LFS) [42] to replace large files is required. Next, the commands to extract the downloaded DB is run.

```
1  git clone ...
        https://github.com/Wireless-Innovation-Forum/SAS-Data.git
2  git lfs install
3  git lfs pull
4  python extract_geo.py
5  python extract_census.py
```

Next, one has to specify the location of the NED in the file *src/harness/* `reference_models`*/geo/CONFIG.py* of the code repository obtained from [12]. Alternatively, it is highly recommended that one would create soft links—Ubuntu command *ln -s*—in the main code repository with *data/geo/ned/* of the downloaded DB pointing to *SAS-Data/ned/data/geo/ned/*. Another option would be to move the extracted files into a directory *data/-geo/ned/* of the main code repository. Ultimately, since the ITM code located at *src/harness/* `reference_models`*/propagation/itm/its/itm.cpp* is a C++ code, one should enable Python to call it. To do so, one can navigate to *src/harness/* `reference_models`*/propagation/itm/* and run the *setup.py* Python code which will generate *itm_its.so* file at *src/harness/* `reference_models`*/propagation/itm/its/* which is the file that Python runs to call the C++ file *itm.cpp*. These are summarized in the script snippet below. Next, Sect. 6.5.2 creates the additional code wrapper to call the ITM propagation and return the pathloss between a TX and a RX.

```
1  cd src/harness/\texttt{reference\_models}/propagation/itm/
2  python setup.py build_ext -i
```

6.5.2   Propagation Code Run

The code wrapper is written below and it should be placed in the directory. This script sets the TX geographic coordinates and AGL height to (30.745948, −81.590209) deg and 10 m, the TX geographic coordinates and AGL height to (30.316543, −81.345152) deg and 1.5 m, the TX operating frequency to 3000 MHz, the TX as outdoor, and reliability of ITM to 0.5, Had the TX been set to indoor, the code would have added 15 dB to the pathloss, currently returned as 186.32859 dB. Also, the reliability of 0.5 represents a median propagation loss for the ITM [21].

```
1   from math import ceil
2   from reference_models.propagation import wf_itm
3   from reference_models.geo import vincenty
4   import numpy as np
5   dTxLat = 30.745948
6   dTxLon = -81.590209
7   dTxH = 10
8   dRxLat = 30.316543
9   dRxLon = -81.345152
10  dRxH = 1.5
11  dFreqMhz = 3000
12  bTX_indoor= False
13  dReliability = 0.5
```

```
14  db_lossItm = wf_itm.CalcItmPropagationLoss(dTxLat, ...
        dTxLon, dTxH, dRxLat, dRxLon, dRxH, reliability= ...
        dReliability, bTX_indoor=False, freq_mhz=dFreqMhz)
15  print db_lossItm
```

6.6 Chapter Summary

This chapter presented the ITM propagation loss calculation by computing the great circle coordinates, their elevation, and climate/refractivity in order to calculate the pathloss. The chapter included algorithms to compute the great circle path based on the Forward and Inverse Vincenty geodesic computations for which algorithms are provided. Then, it included the DBs from which elevation could be extracted and the algorithms to find the relevant DB piece from which to extract the elevation were presented. The chapter subsumed the calculation of climate and refractivity for the great circle path, and portrayed the computation of the ITM pathloss. Finally, the chapter culminated by providing the open source code base, how to obtain the code base, and how to run it.

References

1. H. Friis, A note on a simple transmission formula, in *IRE Proc* (1946)
2. M. Ghorbanzadeh, Resource allocation and end-to-end quality of service for cellular communications systems in congested and contested environments, in *Ph.D. Thesis, Virginia Tech* (2015)
3. M. Ghorbanzadeh, A. Abdelhadi, C. Clacy, *Cellular Communications Systems in Congested Environments Resource Allocation and End-to-End Quality of Service Solutions with MATLAB* (Springer, Berlin, 2017)
4. J. Egli, Radio propagation above 40 mc over irregular terrain, in *Proceedings of the IRE. IEEE* (1957)
5. M. Ghorbanzadeh, E. Visotsky, P. Moorut, W. Yang, C. Clancy, Radar inband and out-of-band interference into LTE macro and small cell uplinks in the 3.5 GHz band, in *Proceedings of the 2015 IEEE Wireless Communications and Networking Conference (WCNC)* (2015)
6. M. Ghorbanzadeh, E. Visotsky, P. Moorut, W. Yang, C. Clancy, Radar in-band interference effects on macrocell LTE uplink deployments in the U.S. 3.5 GHz band, in *Proceedings of the 2015 International Conference on Computing, Networking and Communications (ICNC)* (2015)
7. M. Ghorbanzadeh, E. Visotsky, P. Moorut, W. Yang, C. Clancy, Radar interference into lte base stations in the 3.5 GHz band, in *Elsevier, Physical Communication* (2016)
8. H. Shajaiah, M. Ghorbanzadeh, A. Abdelhadi, C. Clancy, Application-aware resource allocation based on channel information for cellular networks, in *Proceedings of the 2019 IEEE Wireless Communications and Networking Conference (WCNC)* (2019), pp. 1–6
9. M. Ghorbanzadeh, A. Abdelhadi, C. Clancy, Application-aware resource allocation of hybrid traffic in cellular networks. IEEE Trans. Cogn. Commun. Netw. **3**(2), 226–241 (2017)
10. G. Hufford, A. Longley, W. Kissick, A guide to the use of the its irregular terrain model in the area prediction mode, in *US Department of Commerce* (1982)

11. T. Vincenty, Direct and inverse solutions of geodesics on the ellipsoid with application of nested equations. Surv. Rev. **23**(176), (1975)
12. Wireless Innovation Forum (2019). https://github.com/wireless-innovation-forum/spectrum-access-system
13. M. Ghorbanzadeh, Y. Chen, K. Ma, C. Clancy, R. McGwier, A neural network approach to category validation of android applications, in *IEEE Conference on Computing, Networking, and Communications (ICNC)* (2013)
14. M. Ghorbanzadeh, Y. Chen, C. Clancy, Fine-grained end-to-end network model via vector quantization and hidden Markov processes, in *Proceedings of the IEEE Conference on Communications (ICC)* (2013)
15. M. Ghorbanzadeh, A. Abdelhadi, C. Clancy, A utility proportional fairness radio resource block allocation in cellular networks, in *Proceedings of the IEEE International Conference on Computing, Networking and Communications (ICNC)* (2015)
16. M. Ghorbanzadeh, A. Abdelhadi, C. Clancy, A utility proportional fairness bandwidth allocation in radar-coexistent cellular networks, in *Military Communications Conference (MILCOM)* (2014)
17. O.U.G.S. (USGS), National elevation dataset (ned), in *US Geological Survey* (2009)
18. FCC, Use of computer-generated terrain data for determining antenna heights above average terrain, in *FCC 84-341* (1984)
19. ITU-R Recommendation P.2001, *A General Purpose Wide-range Terrestrial Propagation Model in the Frequency Range 30 MHz–50 GHz* (2015)
20. ITU-R Recommendation P.452, *Prediction Procedure for the Evaluation of Interference between Stations on the Surface of the Earth at Frequencies Above about 0.1 GHz, Radio-communication Sector* (2015)
21. G. Hufford, A. Longley, W. Kissick, *A Guide to the Use of the its Irregular Terrain Model in the Area Prediction Mode* (1982)
22. *NTIA ITM Reference Code*. https://www.its.bldrdoc.gov/resources/radio-propagation-software/itm/itm.aspx. Accessed: December 2020
23. P. Rice, A. Longley, K. Norton, A. Barsis, *Transmission Loss Predictions for Tropospheric Communications Circuits* (1978)
24. Python 2.7. https://www.python.org/download/releases/2.7/. Accessed: December 2020
25. Ubuntu for Desktop. https://ubuntu.com/download/desktop. Accessed: December 2020
26. PIP. https://pip.pypa.io/en/stable/installing/. Accessed: December 2020
27. Shapely. http://trac.osgeo.org/geos/. Accessed: December 2020
28. XML for Python. http://lxml.de/installation.html. Accessed: December 2020
29. libgdal for Python. https://pypi.python.org/pypi/gdal/ and http://trac.osgeo.org/gdal/wiki/downloadinggdalbinaries. Accessed: December 2020
30. numpy for Python. http://www.scipy.org/scipylib/download.html. Accessed: December 2020
31. shapely for Python. https://pypi.python.org/pypi/shapely. Accessed: December 2020
32. pyJWT for Python. https://pypi.python.org/pypi/pyjwt. Accessed: December 2020
33. pykml for Python. https://pythonhosted.org/pykml/installation.html. Accessed: December 2020
34. cryptography for Python. https://pypi.python.org/pypi/cryptography. Accessed: December 2020
35. ftputil for Python. http://ftputil.sschwarzer.net/trac/wiki/documentation. Accessed: December 2020
36. json for Python. https://github.com/julian/jsonschema. Accessed: December 2020
37. OpenSSL for Python. https://github.com/pyca/pyopenssl. Accessed: December 2020
38. mock for Python. https://pypi.python.org/pypi/mock. Accessed: December 2020
39. functools32 for Python. https://pypi.python.org/pypi/functools32. Accessed: December 2020
40. psutil for Python. https://github.com/giampaolo/psutil. Accessed: December 2020
41. USGS Database. https://github.com/wireless-innovation-forum/sas-data. Accessed: December 2020
42. LFS. https://git-lfs.github.com/. Accessed: December 2020

Chapter 7
Channel-Aware Resource Allocation Large Scale Simulation

7.1 Introduction

In Chaps. 3 and 4, we introduced a novel convex utility proportional fairness maximization for optimal resource allocation in wireless networks and outfitted the optimization with the subscriber, application status, and service differentiations parameterized, respectively, as UE subscription weights, application status weights, and application utility functions. Over there, we developed a centralized architecture for the proposed resource allocation which assigned application rates by the eNB in a single stage in response to the application utility parameters sent by the UEs to the eNBs. Moreover, we provided with a distributed architecture for the same radio resource allocation framework which was introduced in Chap. 3, which accounted for application types and temporal usages as well as UE priorities, and assigned application rates in two stages from the eNBs to the UEs and by the UEs to the running applications. While we saw the efficacy of the proposed methodology in a real-world WiFi network in Chap. 3, in a realistic large scale wireless communications system, there are other factors to be accounted for. Radio waves undergo various propagation effects [1–7] including path loss, absorption by Oxygen and water vapor [8–14], and diffraction loss, collectively referred to as the channel. When it is said that a UE has a bad channel, it means that based on its current position with respect to its serving BS, the signals transmitted between the BS and the UE and vice versa suffer from severe path loss, and possibly diffraction loss so that the throughput for the UE is much lower. Transmitting under bad condition leads to transmission loss and bit errors, therefore in order to transmit under bad channel conditions, lower order MCS are used. Lower coding MCS means that lower number of bits can be transmitted with one symbol. We provide with a large scale simulation over a large geographic area and show the application of the proposed method. Before proceeding with the simulation, the major points from Chap. 5 are repeated for convenience.

© The Author(s), under exclusive license to Springer Nature Switzerland AG 2022
M. Ghorbanzadeh, A. Abdelhadi, *Practical Channel-Aware Resource Allocation*,
https://doi.org/10.1007/978-3-030-73632-3_7

Table 7.1 Characterizing epsilon for different channels

CQI	M	CR	SE	T1	T2	T3	T4	T5	$SNR_{min,lin}$
0	NT	–	–	–	–	–	–	–	–
1	$QPSK$	0.76	0.1523	1.95	2.00	-7.00	-3.10	-4.80	0.199526231
2	$QPSK$	0.12	0.2344	4.00	4.05	-5.00	-1.15	-2.60	0.316227766
3	$QPSK$	0.19	0.377	6.00	5.10	-3.15	1.50	0.00	0.484172368
4	$QPSK$	0.3	0.6016	8.00	8.00	-1.00	4.00	2.60	0.794328235
5	$QPSK$	0.44	0.877	10.00	10.00	1.00	6.00	4.95	1.258925412
6	$QPSK$	0.59	1.1758	11.95	11.80	3.00	8.90	7.60	1.995262315
7	$16QAM$	0.37	1.4766	14.05	13.90	5.00	12.70	10.60	3.16227766
8	$16QAM$	0.48	1.9141	16.00	16.10	6.90	14.90	12.95	4.897788194
9	$16QAM$	0.6	2.4063	17.90	17.45	8.90	17.50	15.40	7.762471166
10	$64QAM$	0.45	2.7305	19.9	19.50	10.85	20.50	18.10	12.16186001
11	$64QAM$	0.55	3.3223	21.5	21.50	12.60	22.45	20.05	18.19700859
12	$64QAM$	0.65	3.9023	23.45	23.10	14.35	23.20	22.00	27.22701308
13	$64QAM$	0.75	4.5234	25.00	24.90	16.15	24.90	24.55	41.20975191
14	$64QAM$	0.85	5.1152	27.30	27.00	18.15	27.00	26.80	65.31305526
15	$64QAM$	0.95	5.554	22.00	29.10	20.00	29.10	29.60	100.00

M: Modulation, CR: Coding Rate,SE: Spectral Efficiency, T: Transmit Mode [15]

Table 7.1 repeats the LTE DL link adaptation parameters, namely MCS, code rate, and code efficiency, used in Chap. 5 to incorporate the channel effect into the resource allocation optimization of Chap. 3. Moreover, the resource allocation with channel effect present is repeated from Chap. 5 as Eq. (7.1) below whose solutions were, from Chap. 5, in Algorithms 20, 21, and 22. Next, Sect. 7.2 presents the simulation.

$$\max_{\mathbf{k}} \quad \prod_{i=1}^{M} V_i^{\beta_i} (\beta k_i \frac{T_{RB}}{T_f} \frac{n-m}{n} (0.0087 SNR_i + 0.6821) log_2(1 + SNR_i))$$

$$\text{subject to} \quad \sum_{i=1}^{M} k_i \leq K, k_i \geq 0, \quad i = 1, 2, \ldots, M$$

$$(7.1)$$

7.2 Large Scale Network Simulation

Here, we perform a large scale network to observe the effect of resource allocation with channel consideration. We consider a 10 km ×10 km area in Falls Church, Virginia (VA). We consider a cellular system in this area as depicted in Fig. 7.1. The geographic area covers Falls Church and Annandale cities in the Northern VA. The red dots in the picture illustrate the corners of the grid. First, we need to obtain the

---

Algorithm 20 UE channel-aware EURA optimization algorithm

Send initial bid $w_i(1)$ to eNB.
loop
    Receive shadow price $p(n)$ from eNB.
    if STOP from eNB **then**
        Calculate allocated rate $k_i^{\text{opt}} = \frac{w_i(n)}{p(n)}$.
        STOP
    else
        Solve $k_i(n) = \arg\max\limits_{k_i}\Big(\log V_i(\beta k_i \frac{T_{\text{RB}}}{T_f} \frac{n-m}{n} (0.0087 SNR_i + 0.6821) log_2(1 + SNR_i)) -$

        $p(n)k_i \Big)$.
        Send new bid $w_i(n) = p(n)k_i(n)$ to eNB.
    end if
end loop

Algorithm 21 eNB EURA optimization algorithm

---

loop
    Receive bids $w_i(n)$ from UEs. {Let $w_i(0) = 1 \ \forall i$}
    if $|w_i(n) - w_i(n-1)| < \delta \ \forall i$ **then**
        Allocate rates, $k_i^{\text{opt}} = \frac{w_i(n)}{p(n)}$ to user i.
        STOP
    else
        Calculate $p(n) = \frac{\sum_{i=1}^{M} w_i(n)}{K}$.
        Send new shadow price $p(n)$ to all UEs.
    end if
end loop

Algorithm 22 UE IURA algorithm

---

loop
    Receive r_i^{opt} from eNB. {by EURA Algorithms}
    Solve
    $\mathbf{r}_i = \arg\max\limits_{\mathbf{r}_i} \sum_{j=1}^{N_i}(\alpha_{ij} \log U_{ij}(r_{ij}) - p_I r_{ij}) + p_I r_i^{\text{opt}}$ {$\mathbf{r}_i = \{r_{i1}, r_{i2}, \ldots, r_{iN_i}\}$}
    Allocate r_{ij} to the jth application.
end loop

REM all over this geographic area. The REM would include propagation pathloss, diffraction loss, as well as troposcatter loss; however, due to the size of the area, only propagation and diffraction loss would be present [16]. In order to obtain the path loss, we use the ITM model in its point-to-point (P2P) [16] mode. The reason to resort to ITM is the fact that ITM is the only publicly available propagation model that considers morphology and elevation. It is also suitable for a wide range of frequencies from VHF to 20 GHz which makes it accommodate a wide range of applications.

The ITM has two modes of operation, which are the APM and P2P, where the former only accounts for the average changes in the elevation in the geographic area while the latter accounts for the precise elevations on the path. In order to obtain

Fig. 7.1 10 km ×10 km area covering Falls Church and Annandale in Northern VA, where our cell planning and resource allocation occurs

elevation, we download the 1 arc second USGS NED. The data are available for free and provide with elevations with resolution of 30 m. The elevation data are in the form of 32 bit binary files.

Once the REM is obtained, the link budgets in the UL and DL directions are performed in order to obtain SNRs which will be used in the resource allocation algorithms. There will 20 eNBs distributed over the geographic area. And we will be using genetic algorithms to distributed the BSs in the region in a fashion to maximize the coverage of the area. It is noteworthy that the footprint of the BSs is based on the DL SNR 1 dB. Due to the difference of the path loss in various directions, the foot prints will be non-circles. The plot of SNR of the UEs for the eNBs in the system in depicted in Fig. 7.2. As we can see the y-axis (eNB) shows the 20 eNBs in our system and the x axis (UE) shows the UEs that are served by each corresponding

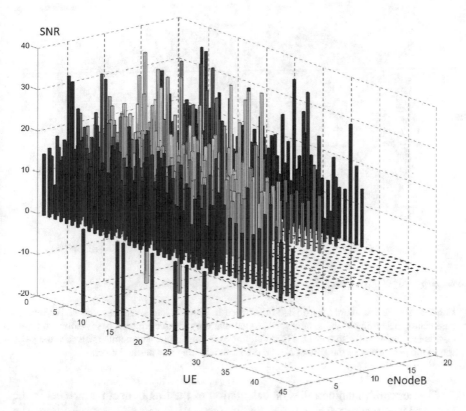

Fig. 7.2 UEs are randomly distributed in the area in Fig. 7.1, and since they choose the serving base station according to the strongest DL signal that they receive from all base stations, they are distributed distinctively amongst the base stations as in this figure. The numbers on each pie piece shows the percentage of the UEs of 500 total UEs that belong to a particular cell

eNB in the y-axis. Since, UEs chose their serving eNB based on the received SNR, the number of UEs in the different eNBs are different as well. Furthermore, we can see some UEs with very low SNRs (such as those in the eNB 1). This is because we distributed the UEs purely randomly, and therefore, some UEs are in bad channel conditions.

Once the eNBs are distributed all over the area, then we randomly distribute the UEs all over the grid and use the DL SNR from the BSs to help UEs pick their serving BSs. As such, we do not drop the UEs under BSs footprint, instead, we distribute them at random and the SNRs define the serving BS. Then, the UL SNR is leveraged to obtain the coefficients which are used in Eq. (5.8). We are assuming omnidirectional antennae and no interference is accounted for. The UEs choose their serving BS based on the strongest signal that they receive from each BS in the DL direction. Therefore, the UEs will be distributed in various cells of the BSs as in Fig. 7.3.

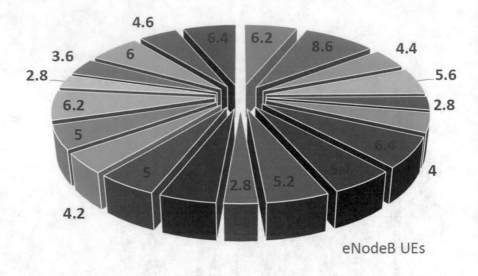

eNodeB UEs

■ 1 ■ 2 ■ 3 ■ 4 ■ 5 ■ 6 ■ 7 ■ 8 ■ 9 ■ 10 ■ 11 ■ 12 ■ 13 ■ 14 ■ 15 ■ 16 ■ 17 ■ 18 ■ 19 ■ 20

Fig. 7.3 UEs are randomly distributed in the area in Fig. 7.1, and since they choose the serving base station according to the strongest DL signal that they receive from all base stations, they are distributed distinctively amongst the base stations as in this figure. The numbers on each pie piece shows the percentage of the UEs of 500 total UEs that belong to a particular cell

The network planning relies on calculation of the link budget parameters in the UP and DL directions. The UL link budget has a fixed part and a varying part. The first part can be calculated as Eq. (7.2). Here, P_i is the ith UE effective radiated isotropically radiated power (EIRP), which is the sum of the UE transmit power in dB and the gain in the direction of the eNB in dBi. The L_{Cable} is the cable loss in dB, the I is the interference margin in dB, G_{TX} is the eNB antenna gain, and G_{MH} is eNB master-head gain [15]. In the simulation, we take the UE EIRP as 24 dBm, interference margin as 2 dB, cable loss as 1 dB, eNB antenna gain as 6 dBi, and master-head gain as 2 dB.

$$P_i^{UL} = P_{UE} - I + L_{Cable} + G_{TX} + G_{MH} - L_{Implementation\ Margin}$$
$$- L_{Fast\ Fading\ Margin} - L_{Body\ Loss} + G_{RX} - L_{Loss} \tag{7.2}$$

Furthermore, in Eq. (7.2), we take the implementation margin 5 dB, body loss 3 dB, receiver gain 6 dB, fast fading margin 5 dB, and clutter loss 12 dB. In is noteworthy that the 12 dB clutter loss is conservative, and a precise clutter loss can be obtained using USGS Land Use Land Cover (LCLU) data [17]. However, this is out of the scope of this simulation. Besides, the noise figure (NF) for the receiver, i.e. eNB in the UL direction, is 2 dB, which accumulates to receiver noise floor density as Eq. (7.3). The second term and third term in the addition comes from the thermal noise, evaluated at 10 MHz LTE, and at the temperature 290 degrees of Kelvin (K).

$$ND_{eNB} = NF_{eNB} + 30 + 10\log(1.38e^{-23} \times 290) \tag{7.3}$$

Then, the signal strength at the jth eNB position can be calculated as Eq. (7.4), where $P_{i,j}$ is the signal strength from the ith UE to the jth eNB, whose noise floor density is ND_j, and $L_{REM}(i, j)$ is the pathloss between the ith UE to the jth eNB.

$$P_{i,j}^{UL} = P_i^{UL} - L_{REM}(i, j) - ND_j - 10\log(180 \times 1000) \tag{7.4}$$

Furthermore, in the DL direction, the link budget can be calculated as Eq. (7.5).

$$P_i^{DL} = P_{eNB} - I + L_{Cable} + G_{MH} - L_{Implementation\ Margin} \tag{7.5}$$
$$-L_{Fast\ Fading\ Margin} - L_{Body\ Loss} + G_{RX} - L_{Loss}$$

Then, the signal strength at the jth UE position can be calculated as Eq. (7.6), where $P_{i,j}$ is the signal strength from the ith eNB to the jth UE, whose noise floor density is ND_j, and $L_{REM}(i, j)$ is the pathloss between the ith eNB to the jth UE.

$$P_{i,j}^{DL} = P_i^{DL} - L_{REM}(i, j) - ND_j - 10\log(180 \times 1000) \tag{7.6}$$

For eNB 1, the plots of resource elements allocated to the UEs as well as the throughput in Mbps are given in Fig. 7.4a,b, respectively. As we can observe from the figures, the horizontal axis is the UE indices which indicate that there are 31 UEs being served by the eNB 1. The vertical axis is the number of resource elements allocated by the optimization in Algorithms 12 and 13. Each UE has only 1 application running. This assumption simplifies the simulation because the goal of this chapter is observing the effect of the channel for which channel-aware EURA is performed. On the other hand, inside the UE there is no channel and the IURA is not needed to show any channel effect but mere allocation to the applications, which was presented extensively in Chaps. 3 and 4. Furthermore, the legend shows the SNR of the UEs in dB. The bit rate requirements of the applications are according to $\{0.25, 1, 5, 0.25, 1, 5, \ldots\}$ Mbps.

As we can observe from Fig. 7.4b, the UEs with low SNRs are receiving more resource elements in order to meet their bit rate requirements. On the other hand, UEs with high SNR are receiving less resources. On the other hand, Fig. 7.4a shows that the throughputs of UEs 8, 14, 15, 24 are not met. This is due to the -20 dB SNR which is placed on the network by UEs 8, 14, 15, 20, 24, 26, and 29 each with 1, 1, 5, 1, 5, 1, and 1 Mbps which are high rates at low SNRs and indicate using many REs. We can see these in Fig. 7.4b where there is a spike at UEs 8, 14, 15, 20, 24, 26, and 29 which shows the algorithm is allocating more REs to these UEs.

Moreover, as we can see from Fig. 7.5, UEs 8, 14, 15, 20, 24, 26, and 29 are bidding the highest due to the fact that they require more resources in view of their bad channel conditions. This plot shows the last iteration of the algorithm where the shadow prices are converged. Furthermore, we can observe the coverage area of eNB 1 in Fig. 7.6a, and as we see the coverage is not circular which is a

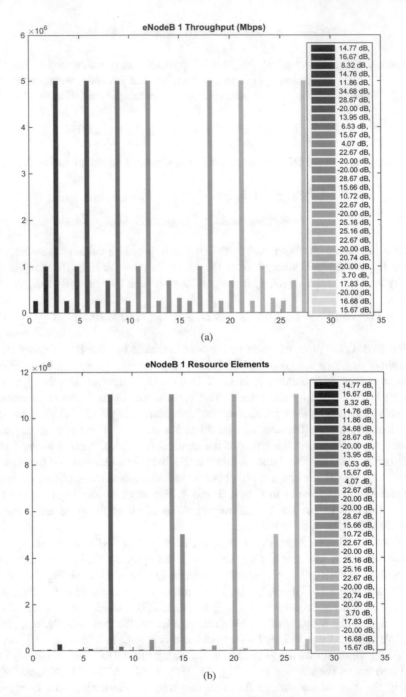

Fig. 7.4 The system contains 31 UEs, each concurrently running a real-time application. (**a**) Throughput of the UEs for eNB 1. (**b**) Resources Allocated to UEs by eNB 1

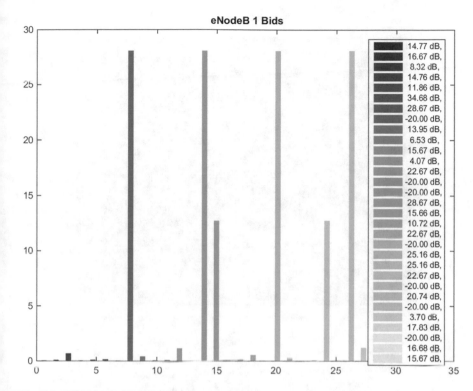

Fig. 7.5 UE Bids pledged to eNB 1

result of REM and elevations in various directions leading to different SNRs. The black addition symbol in the middle of the coverage, shown in yellow, is the eNB 1 coordinates in the 101×101 grid. The side bar shows 0—dark blue—and 1—yellow—for the areas which are, respectively, not covered and are under coverage. The green dots in the region are the UEs scattered all over the area. As we can see, some of the UEs are not in the foot print of the eNBs, and this is due to the fact that the UE choose their serving eNBs only based on the received SNR and as a result, many of the were randomly distributed in the area and were under bad channel conditions (especially those at -20 dB as we discussed). These simply chose eNB1 as it was the strongest signal they could ever receive. Furthermore, we can see the REM which shows that a blue dot at coordinate $(50, 83)$, which is for the eNB 1 and the blue color from the side bar represents a 0 dB loss which is expected as we are at the eNB position. However, as we go further away from the eNB, the path loss increases, which is shown by the color spectrum light blue, green, and yellow in the order of increase. The highest pathloss in the area is about 215 dB.

For eNB 2, the plots of resource elements allocated to the UEs as well as the throughput in Mbps are given in Fig. 7.7a,b, respectively. As we can observe from the figures, the horizontal axis is the UE indices which indicate that there are

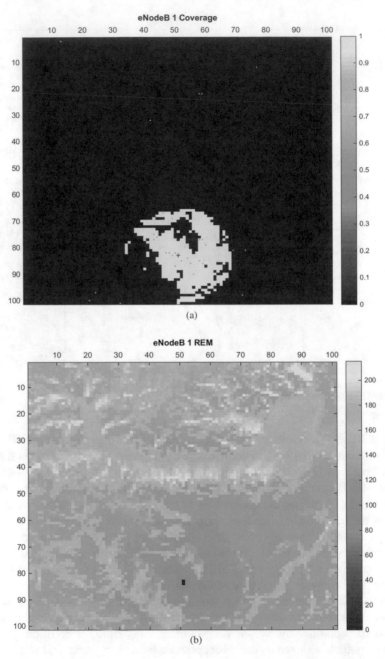

Fig. 7.6 The system contains 31 UEs, each concurrently running a real-time application. (**a**) eNB 1 Coverage. (**b**) eNB 1 REM

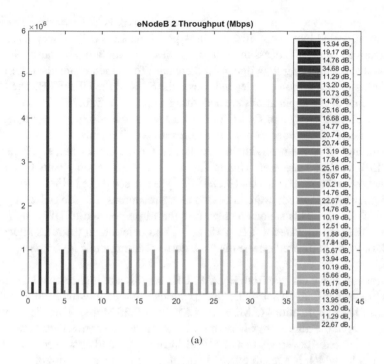

(a)

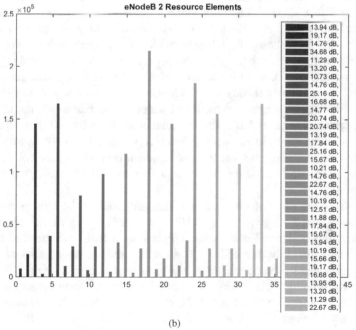

(b)

Fig. 7.7 The system contains 23 UEs, each concurrently running a real-time application. (**a**) Throughput of the UEs for eNB 2. (**b**) Resources Allocated to UEs by eNB 2

23 UEs being served by the eNB 2. The vertical axis is the number of resource elements allocated by the optimization in Algorithms 12 and 13. Each UE has only 1 application running. This assumption simplifies the simulation because the goal of this chapter is observing the effect of the channel for which channel-aware EURA is performed. Furthermore, the legend shows the SNR of the UEs in dB. The bit rate requirements of the applications are according to $\{0.25, 1, 5, 0.25, 1, 5, \ldots\}$ Mbps.

As we can observe from Fig. 7.7b, the UEs with low SNRs are receiving more resource elements in order to meet their bit rate requirements. On the other hand, UEs with high SNR are receiving less resources. On the other hand, Fig. 7.4a shows that UE throughputs are met. This is due all UEs are at good channel conditions such that the minimum SNR was 10.1885 dB. We can see these in Fig. 7.7b that UEs with higher bit rate needs were allocated more resources, i.e. for UE k, $R_{1+3K} < R_{2+3K} < R_{3+3K}$. This statement says that the rate allocated to UE 1, 4, 7, $\ldots$ is less than the rate allocated to UEs 2, 5, 8, $\ldots$ and is less than the rates allocated to UEs 3, 6, 9, $\ldots$ compared one to one (i.e. corresponding indices compared to each other).

Moreover, as we can see from Fig. 7.8, UEs $3 + 3K$ are higher than UEs $2 + 3K$ which bid higher than UEs $1 + 3K$ due to the fact that they require more resources in view of their applications (5 Mbps vs. 1 Mbps vs 0.25 Mbps). This plot shows the last iteration of the algorithm where the shadow prices are converged. Furthermore, we can observe the coverage area of eNB 1 in Fig. 7.9a, and as we see the coverage is not circular which is a result of REM and elevations in various directions leading to different SNRs. The black addition symbol in the middle of the coverage, shown in yellow, is the eNB 2 coordinates in the 101×101 grid. The side bar shows 0— dark blue—and 1—yellow—for the areas which are, respectively, not covered and are under coverage. The green dots in the region are the UEs scattered all over the area. As we can see, UEs are in the foot print of the eNBs and had good channel conditions. These simply chose eNB 2 as it was the strongest signal they could ever receive. Furthermore, we can see the REM which shows that a blue dot at the eNB location, which is for the eNB 2 and the blue color from the side bar represents a 0 dB loss which is expected as we are at the eNB position. However, as we go further away from the eNB, the path loss increases, which is shown by the color spectrum light blue, green, and yellow in the order of increase. The highest pathloss in the area is about 210 dB.

For eNB 3, the plots of resource elements allocated to the UEs as well as the throughput in Mbps are given in Fig. 7.10a,b, respectively. As we can observe from the figures, the horizontal axis is the UE indices which indicate that there are 22 UEs being served by the eNB 3. The vertical axis is the number of resource elements allocated by the optimization in Algorithms 12 and 13. Each UE has only 1 application running. This assumption simplifies the simulation because the goal of this chapter is observing the effect of the channel for which channel-aware EURA is performed. Furthermore, the legend shows the SNR of the UEs in dB. The bit rate requirements of the applications are according to $\{0.25, 1, 5, 0.25, 1, 5, \ldots\}$ Mbps.

As we can observe from Fig. 7.10b, the UEs with low SNRs are receiving more resource elements in order to meet their bit rate requirements. On the other hand,

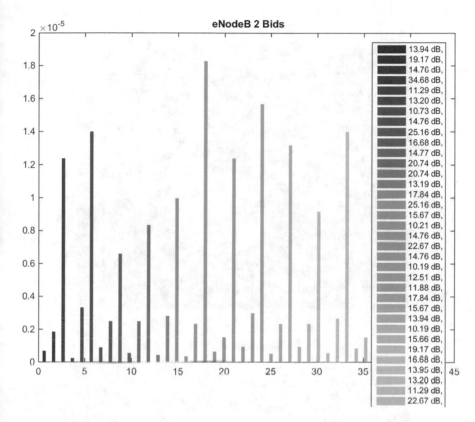

Fig. 7.8 UE Bids pledged to eNB 2

UEs with high SNR are receiving less resources. On the other hand, Fig. 7.10a shows that UE throughputs are met. This is due all UEs are at good channel conditions such that the minimum SNR was 10.1885 dB. We can see these in Fig. 7.7b that UEs with higher bit rate needs were allocated more resources, i.e. for UE k, $R_{1+3K} < R_{2+3K} < R_{3+3K}$. This statement says that the rate allocated to UE 1, 4, 7, ... is less than the rate allocated to UEs 2, 5, 8, ... and is less than the rates allocated to UEs 3, 6, 9, ... compared one to one (i.e. corresponding indices compared to each other). Also, we see that for same bit rate requirements, UEs with higher SNR are allocated less resources. For instance, UEs 6, 15, and 18 are allocated less REs than UEs 3, 9, and 12 even though the bit rate requirements for both are 5 Mbps. This is due to the fact the these UEs have higher SNRs. Moreover, amongst UEs 6, 15, and 18, the last one has the smallest SNR 15.67 dB as opposed to the 19.71 SNR for UE 15 which causes UE 18 to get more resources as opposed to UE 15. On the other hand, UE 12 has the lowest SNR in the system, 12.97 dB while it is a high bit requirement application (5 Mbps), so it needs more REs as it is assigned by the algorithm.

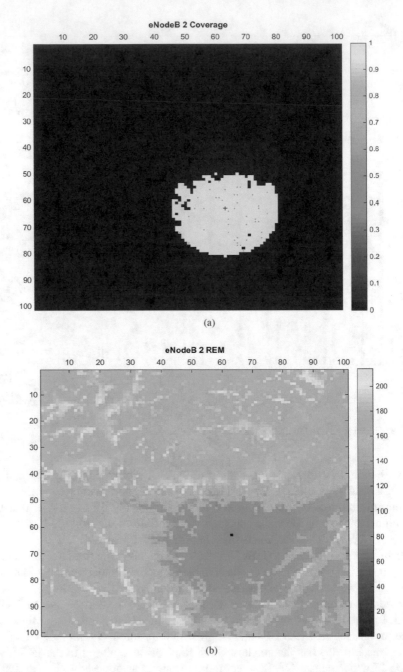

Fig. 7.9 The system contains 23 UEs, each concurrently running a real-time application. (**a**) eNB 2 Coverage. (**b**) eNB 2 REM

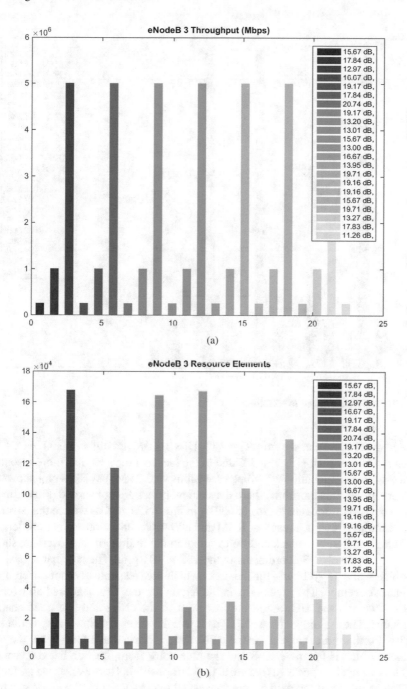

Fig. 7.10 The system contains 22 UEs, each concurrently running a real-time application. (**a**) Throughput of the UEs for eNB 3. (**b**) Resources Allocated to UEs by eNB 3

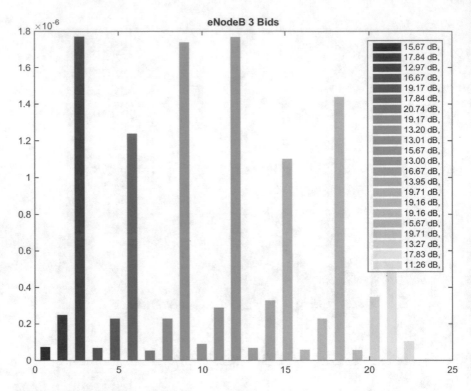

Fig. 7.11 UE Bids pledged to eNB 3

Moreover, as we can see from Fig. 7.11, UEs $3 + 3K$ are higher than UEs $2 + 3K$ which bid higher than UEs $1 + 3K$ due to the fact that they require more resources in view of their applications (5 Mbps vs. 1 Mbps vs 0.25 Mbps). This plot shows the last iteration of the algorithm where the shadow prices are converged. Furthermore, we can observe the coverage area of eNB 1 in Fig. 7.12a, and as we see the coverage is not circular which is a result of REM and elevations in various directions leading to different SNRs. The black addition symbol in the middle of the coverage, shown in yellow, is the eNB 3 coordinates in the 101×101 grid. The side bar shows 0—dark blue—and 1—yellow—for the areas which are, respectively, not covered and are under coverage. The green dots in the region are the UEs scattered all over the area. As we can see, UEs are in the foot print of the eNBs and had good channel conditions. These simply chose eNB 3 as it was the strongest signal they could ever receive. Furthermore, we can see the REM which shows that a blue dot at the eNB location, which is for the eNB 3 and the blue color from the side bar represents a 0 dB loss which is expected as we are at the eNB position. However, as we go further away from the eNB, the path loss increases, which is shown by the color spectrum light blue, green, and yellow in the order of increase. The highest pathloss in the area is about 218 dB.

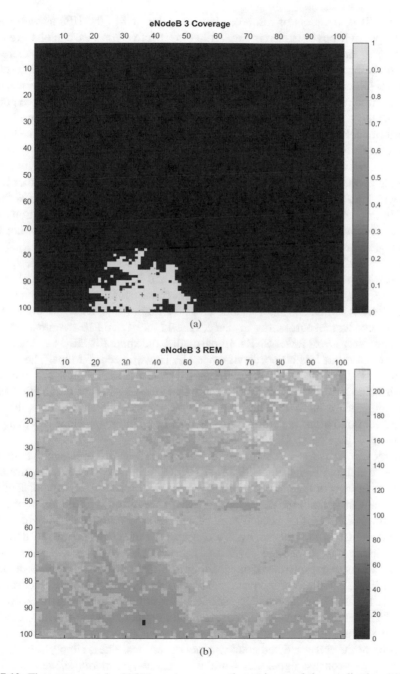

Fig. 7.12 The system contains 22 UEs, each concurrently running a real-time application. (**a**) eNB 3 Coverage. (**b**) eNB 3 REM

For eNB 4, the plots of resource elements allocated to the UEs as well as the throughput in Mbps are given in Fig. 7.13a,b, respectively. As we can observe from the figures, the horizontal axis is the UE indices which indicate that there are 28 UEs being served by the eNB 4. The vertical axis is the number of resource elements allocated by the optimization in Algorithms 12 and 13. Each UE has only 1 application running. This assumption simplifies the simulation because the goal of this chapter is observing the effect of the channel for which channel-aware EURA is performed. Furthermore, the legend shows the SNR of the UEs in dB. The bit rate requirements of the applications are according to {0.25, 1, 5, 0.25, 1, 5, ...} Mbps.

As we can observe from Fig. 7.13b, the UEs with low SNRs are receiving more resource elements in order to meet their bit rate requirements. On the other hand, UEs with high SNR are receiving less resources. On the other hand, Fig. 7.13a shows that UE throughputs are met. This is due all UEs are at good channel conditions such that the minimum SNR was 10.1885 dB. We can see these in Fig. 7.13b that UEs with higher bit rate needs were allocated more resources, i.e. for UE k, $R_{1+3K} < R_{2+3K} < R_{3+3K}$. This statement says that the rate allocated to UE 1, 4, 7, ... is less than the rate allocated to UEs 2, 5, 8, ... and is less than the rates allocated to UEs 3, 6, 9, ... compared one to one (i.e. corresponding indices compared to each other). Also, we see that for same bit rate requirements, UEs with higher SNR are allocated less resources. For instance, UEs 11 and 12 and 18 are allocated more REs since they are at lower SNRs. In particular, the spike for UE 11 is interesting even though it has less bit needs vis-a-vis UE 12 who needs 5 Mbps. The spike is because this UE is at the lowest SNR situation in the system 1.58 dB. We see a similar behavior for UE 20 which is getting more REs vs other UEs with 1 Mbps requirement since its SNR is lower. On the other hand, UE 15 who has a high bit requirement of 5 Mbps is getting less REs than its counterparts since it is at a good channel condition 25.16 dB.

Moreover, as we can see from Fig. 7.14, UEs $3 + 3K$ are higher than UEs $2 + 3K$ which bid higher than UEs $1 + 3K$ due to the fact that they require more resources in view of their applications (5 Mbps vs. 1 Mbps vs 0.25 Mbps). This plot shows the last iteration of the algorithm where the shadow prices are converged. Interestingly, UE 15 that needed high resources at good channel is bidding less than its counterparts. Furthermore, we can observe the coverage area of eNB 4 in Fig. 7.15a, and as we see the coverage is not circular which is a result of REM and elevations in various directions leading to different SNRs. The black addition symbol in the middle of the coverage, shown in yellow, is the eNB 4 coordinates in the 101×101 grid. The side bar shows 0—dark blue—and 1—yellow—for the areas which are, respectively, not covered and are under coverage. The green dots in the region are the UEs scattered all over the area. As we can see, UEs are in the foot print of the eNBs and had good channel conditions. These simply chose eNB 4 as it was the strongest signal they could ever receive. Furthermore, we can see the REM which shows that a blue dot at the eNB location, which is for the eNB 3 and the blue color from the side bar represents a 0 dB loss which is expected as we are at the eNB position. However, as we go further away from the eNB, the path loss

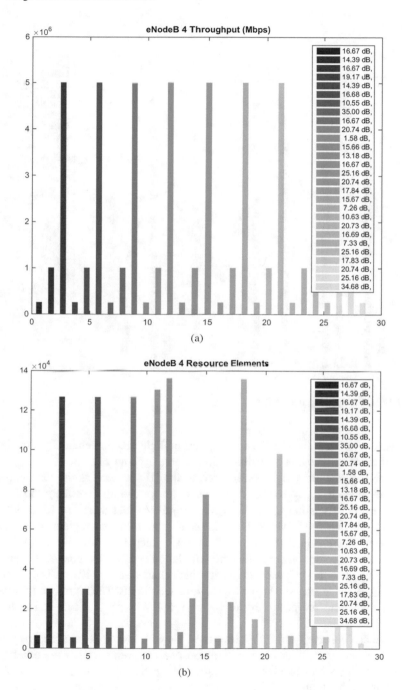

Fig. 7.13 The system contains 28 UEs, each concurrently running a real-time application. (**a**) Throughput of the UEs for eNB 4. (**b**) Resources Allocated to UEs by eNB 4

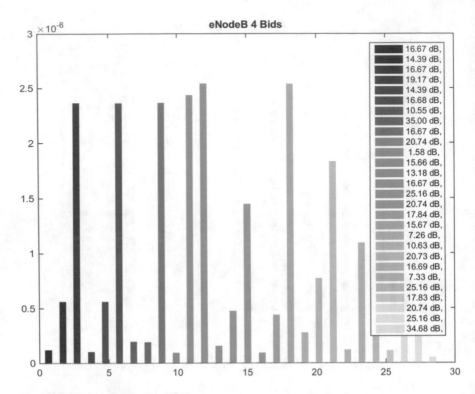

Fig. 7.14 UE Bids pledged to eNB 4

increases, which is shown by the color spectrum light blue, green, and yellow in the order of increase. The highest pathloss in the area is about 211 dB.

For eNB 5, the plots of REs allocated to the UEs as well as the throughput in Mbps are given in Fig. 7.16a,b, respectively. As we can observe from the figures, the horizontal axis is the UE indices which indicate that there are 14 UEs being served by the eNB 5. The vertical axis is the number of resource elements allocated by the optimization in Algorithms 12 and 13. Each UE has only 1 application running. This assumption simplifies the simulation because the goal of this chapter is observing the effect of the channel for which channel-aware EURA is performed. Furthermore, the legend shows the SNR of the UEs in dB. The bit rate requirements of the applications are according to $\{0.25, 1, 5, 0.25, 1, 5, \ldots\}$ Mbps. The lowest SNR is 9.6584 dB.

As we can observe from Fig. 7.16b, the UEs with low SNRs are receiving more resource elements in order to meet their bit rate requirements. On the other hand, UEs with high SNR are receiving less resources. On the other hand, Fig. 7.16a shows that UE throughputs are met. This is due all UEs are at good channel conditions such that the minimum SNR was 9.6584 dB. We can see these in Fig. 7.16b that UEs with higher bit rate needs were allocated more resources, i.e. for UE k, $R_{1+3K} < R_{2+3K} < R_{3+3K}$. This statement says that the rate allocated to UE 1, 4, 7, ... is

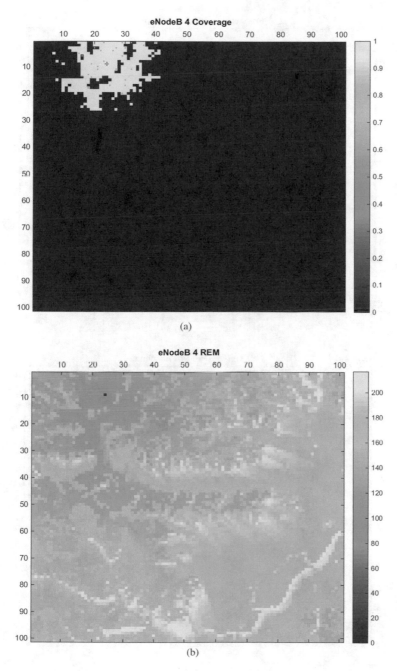

Fig. 7.15 The system contains 28 UEs, each concurrently running a real-time application. (**a**) eNB 4 Coverage. (**b**) eNB 4 REM

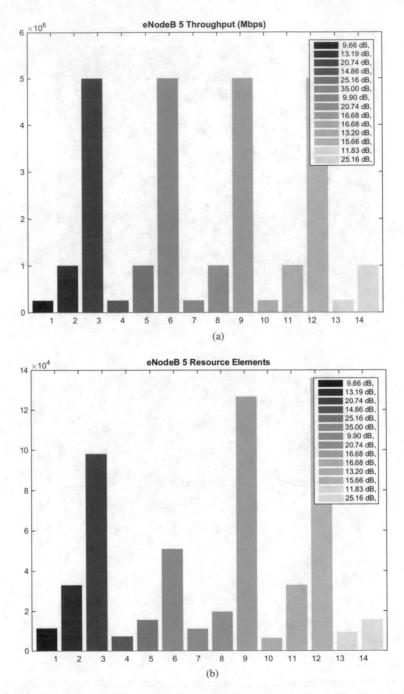

Fig. 7.16 The system contains 14 UEs, each concurrently running a real-time application. (**a**) Throughput of the UEs for eNB 5. (**b**) Resources Allocated to UEs by eNB 5

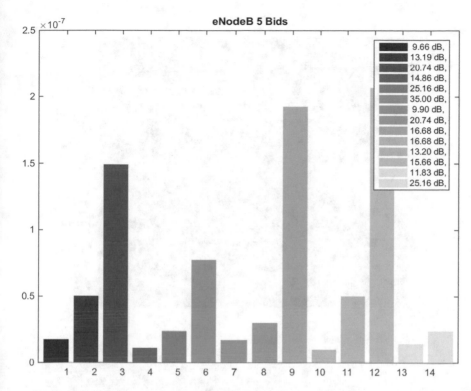

Fig. 7.17 UE Bids pledged to eNB 5

less than the rate allocated to UEs 2, 5, 8, ... and is less than the rates allocated to UEs 3, 6, 9, ... compared one to one (i.e. corresponding indices compared to each other). Also, we see that for same bit rate requirements, UEs with higher SNR are allocated less resources. For instance, UEs 5 is allocated more REs than UEs 2, 8, 11, and 14 since it has a better channel condition.

Moreover, as we can see from Fig. 7.17, UEs $3 + 3K$ are higher than UEs $2 + 3K$ which bid higher than UEs $1 + 3K$ due to the fact that they require more resources in view of their applications (5 Mbps vs. 1 Mbps vs 0.25 Mbps). This plot shows the last iteration of the algorithm where the shadow prices are converged. Furthermore, we can observe the coverage area of eNB 4 in Fig. 7.18a, and as we see the coverage is not circular which is a result of REM and elevations in various directions leading to different SNRs. The black addition symbol in the middle of the coverage, shown in yellow, is the eNB 5 coordinates in the 101×101 grid. The side bar shows 0—dark blue—and 1—yellow—for the areas which are, respectively, not covered and are under coverage. The green dots in the region are the UEs scattered all over the area. As we can see, UEs are in the foot print of the eNBs and had good channel conditions. These simply chose eNB 5 as it was the strongest signal they could ever receive. Furthermore, we can see the REM which shows that a blue dot at the eNB

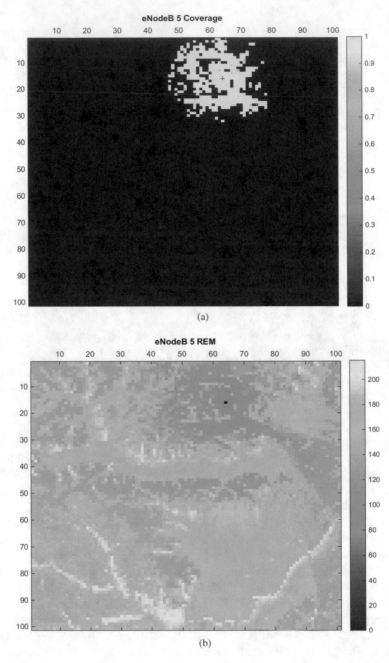

Fig. 7.18 The system contains 14 UEs, each concurrently running a real-time application. (**a**) eNB 5 Coverage. (**b**) eNB 5 REM

location, which is for the eNB 5 and the blue color from the side bar represents a
0 dB loss which is expected as we are at the eNB position. However, as we go further
away from the eNB, the path loss increases, which is shown by the color spectrum
light blue, green, and yellow in the order of increase. The highest pathloss in the
area is about 210 dB.

For eNB 6, the plots of resource elements allocated to the UEs as well as the
throughput in Mbps are given in Fig. 7.19a,b, respectively. As we can observe from
the figures, the horizontal axis is the UE indices which indicate that there are
28 UEs being served by the eNB 4. The vertical axis is the number of resource
elements allocated by the optimization in Algorithms 12 and 13. Each UE has only
1 application running. This assumption simplifies the simulation because the goal of
this chapter is observing the effect of the channel for which channel-aware EURA
is performed. Furthermore, the legend shows the SNR of the UEs in dB. The bit rate
requirements of the applications are according to $\{0.25, 1, 5, 0.25, 1, 5, \ldots\}$ Mbps.

As we can observe from Fig. 7.19b, the UEs with low SNRs are receiving more
resource elements in order to meet their bit rate requirements. On the other hand,
UEs with high SNR are receiving less resources. On the other hand, Fig. 7.19a shows
that UE throughputs are met. This is due all UEs are at good channel conditions
such that the minimum SNR was 9.65 dB. We can see these in Fig. 7.19b that UEs
with higher bit rate needs were allocated more resources, i.e. for UE k, $R_{1+3K} <
R_{2+3K} < R_{3+3K}$. This statement says that the rate allocated to UE 1, 4, 7, ... is
less than the rate allocated to UEs 2, 5, 8, ... and is less than the rates allocated to
UEs 3, 6, 9, ... compared one to one (i.e. corresponding indices compared to each
other). Also, we see that for same bit rate requirements, UEs with higher SNR are
allocated less resources. For instance, UEs 9 and 15 are allocated more REs than
UEs 3 and 5 since they are at lower SNRs.

Moreover, as we can see from Fig. 7.20, UEs $3 + 3K$ are higher than UEs
$2 + 3K$ which bid higher than UEs $1 + 3K$ due to the fact that they require more
resources in view of their applications (5 Mbps vs. 1 Mbps vs 0.25 Mbps). This plot
shows the last iteration of the algorithm where the shadow prices are converged.
Interestingly, UE 15 that needed high resources at good channel is bidding less
than its counterparts. Furthermore, we can observe the coverage area of eNB 6 in
Fig. 7.21a, and as we see the coverage is not circular which is a result of REM
and elevations in various directions leading to different SNRs. The black addition
symbol in the middle of the coverage, shown in yellow, is the eNB 6 coordinates
in the 101×101 grid. The side bar shows 0—dark blue—and 1—yellow—for the
areas which are, respectively, not covered and are under coverage. The green dots
in the region are the UEs scattered all over the area. As we can see, UEs are in the
foot print of the eNBs and had good channel conditions. These simply chose eNB 6
as it was the strongest signal they could ever receive. Furthermore, we can see the
REM which shows that a blue dot at the eNB location, which is for the eNB 6 and
the blue color from the side bar represents a 0 dB loss which is expected as we are
at the eNB position. However, as we go further away from the eNB, the path loss
increases, which is shown by the color spectrum light blue, green, and yellow in the
order of increase. The highest pathloss in the area is about 216 dB.

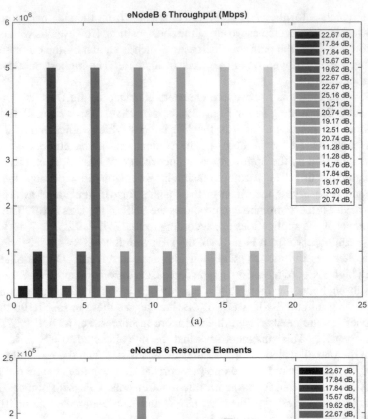

(a)

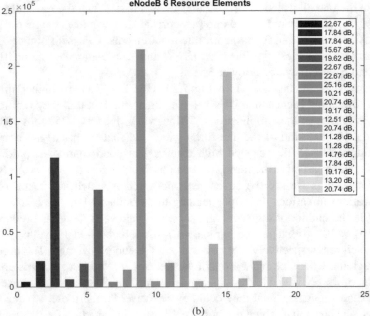

(b)

Fig. 7.19 The system contains 20 UEs, each concurrently running a real-time application. (**a**) Throughput of the UEs for eNB 6. (**b**) Resources Allocated to UEs by eNB 6

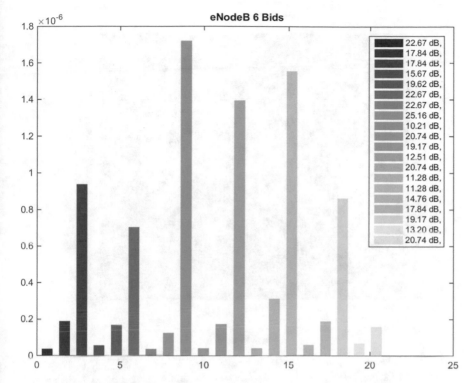

Fig. 7.20 UE Bids pledged to eNB 6

For eNB 7, the plots of resource elements allocated to the UEs as well as the throughput in Mbps are given in Fig. 7.22a,b, respectively. As we can observe from the figures, the horizontal axis is the UE indices which indicate that there are 32 UEs being served by the eNB 7. The vertical axis is the number of resource elements allocated by the optimization in Algorithms 12 and 13. Each UE has only 1 application running. This assumption simplifies the simulation because the goal of this chapter is observing the effect of the channel for which channel-aware EURA is performed. Furthermore, the legend shows the SNR of the UEs in dB. The bit rate requirements of the applications are according to {0.25, 1, 5, 0.25, 1, 5, . . .} Mbps.

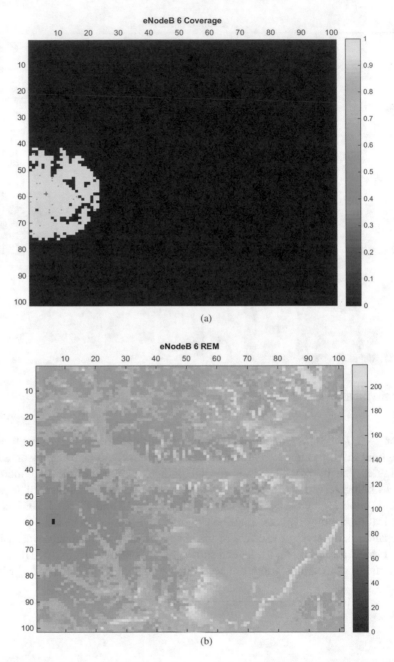

Fig. 7.21 The system contains 20 UEs, each concurrently running a real-time application. (**a**) eNB 6 Coverage. (**b**) eNB 6 REM

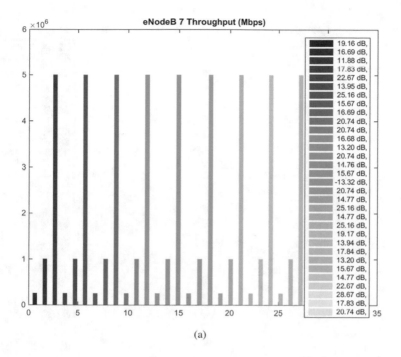

(a)

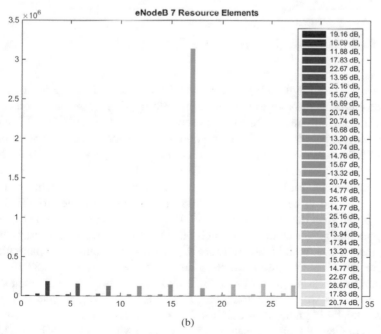

(b)

Fig. 7.22 The system contains 32 UEs, each concurrently running a real-time application. (**a**) Throughput of the UEs for eNB 7. (**b**) Resources Allocated to UEs by eNB 7

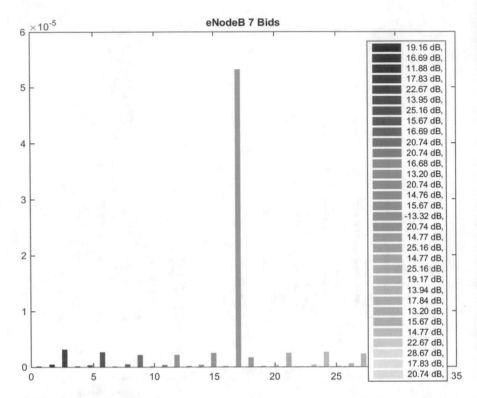

Fig. 7.23 UE Bids pledged to eNB 7

As we can observe from Fig. 7.22b, the UEs with low SNRs are receiving more resource elements in order to meet their bit rate requirements. On the other hand, UEs with high SNR are receiving less resources. On the other hand, Fig. 7.22a shows that UE throughputs are met. This is due all UEs are at good channel conditions such that all SNRs were two digit positive and the only bad SNR was -13.31 dB for UE 17. We can see these in Fig. 7.22b that UEs with higher bit rate needs were allocated more resources, i.e. for UE k, $R_{1+3K} < R_{2+3K} < R_{3+3K}$. This statement says that the rate allocated to UE 1, 4, 7, ... is less than the rate allocated to UEs 2, 5, 8, ... and is less than the rates allocated to UEs 3, 6, 9, ... compared one to one (i.e. corresponding indices compared to each other). Also, we see that for same bit rate requirements, UEs with higher SNR are allocated less resources. In particular, the spike for UE 17 is interesting which is due to its low SNR at -13.32 dB, so the algorithm has to assign more REs to this UE to meet its bit rate requirements.

Moreover, as we can see from Fig. 7.23, UEs $3 + 3K$ are higher than UEs $2 + 3K$ which bid higher than UEs $1 + 3K$ due to the fact that they require more resources in view of their applications (5 Mbps vs. 1 Mbps vs 0.25 Mbps). This plot shows the last iteration of the algorithm where the shadow prices are converged. Interestingly, UE 15 that needed high resources at good channel is bidding less than its counterparts. Furthermore, we can observe the coverage area of eNB 4 in Fig. 7.24a, and as we see the coverage is not circular which is a result of REM and elevations in various directions leading to different SNRs. The black addition symbol in the middle of the coverage, shown in yellow, is the eNB 7 coordinates in the 101 × 101 grid. The side bar shows 0—dark blue—and 1—yellow—for the areas which are, respectively, not covered and are under coverage. The green dots in the region are the UEs scattered all over the area. As we can see, UEs are in the foot print of the eNBs and had good channel conditions. These simply chose eNB 7 as it was the strongest signal they could ever receive. Furthermore, we can see the REM which shows that a blue dot at the eNB location, which is for the eNB 7 and the blue color from the side bar represents a 0 dB loss which is expected as we are at the eNB position. However, as we go further away from the eNB, the path loss increases, which is shown by the color spectrum light blue, green, and yellow in the order of increase. The highest pathloss in the area is about 214 dB.

For eNB 8, the plots of REs allocated to the UEs as well as the throughput in Mbps are given in Fig. 7.25a,b, respectively. As we can observe from the figures, the horizontal axis is the UE indices which indicate that there are 27 UEs being served by the eNB 8. The vertical axis is the number of resource elements allocated by the optimization in Algorithms 12 and 13. Each UE has only 1 application running. This assumption simplifies the simulation because the goal of this chapter is observing the effect of the channel for which channel-aware EURA is performed. Furthermore, the legend shows the SNR of the UEs in dB. The bit rate requirements of the applications are according to $\{0.25, 1, 5, 0.25, 1, 5, \ldots\}$ Mbps.

As we can observe from Fig. 7.25b, the UEs with low SNRs are receiving more resource elements in order to meet their bit rate requirements. On the other hand, UEs with high SNR are receiving less resources. On the other hand, Fig. 7.25a shows that UE throughputs are met. This is due all UEs are at good channel conditions such that the minimum SNR was -15.4759 dB. We can see these in Fig. 7.25b that UEs with higher bit rate needs were allocated more resources, i.e. for UE k, $R_{1+3K} < R_{2+3K} < R_{3+3K}$. This statement says that the rate allocated to UE 1, 4, 7, . . . is less than the rate allocated to UEs 2, 5, 8, . . . and is less than the rates allocated to UEs 3, 6, 9, . . . compared one to one (i.e. corresponding indices compared to each other). Also, we see that for same bit rate requirements, UEs with higher SNR are allocated less resources. For instance, UEs 10 and 26 are allocated more REs since they are at lower SNRs -14.80 and -15.48 dB, respectively.

Also, as we can see from Fig. 7.26, UEs $3+3K$ are higher than UEs $2+3K$ which bid higher than UEs $1+3K$ due to the fact that they require more resources in view of their applications (5 Mbps vs. 1 Mbps vs 0.25 Mbps). This plot shows the last iteration of the algorithm where the shadow prices are converged. Interestingly, UE 15 that needed high resources at good channel is bidding less than its counterparts. Furthermore, we can observe the coverage area of eNodeB 8 in Fig. 7.27a, and as we see the coverage is not circular which is a result of REM and elevations in various directions leading to different SNRs. The black addition symbol in the middle of the coverage, shown in yellow, is the eNB 8 coordinates in the 101 × 101 grid. The side bar shows 0—dark blue—and 1—yellow—for the areas which are, respectively, not

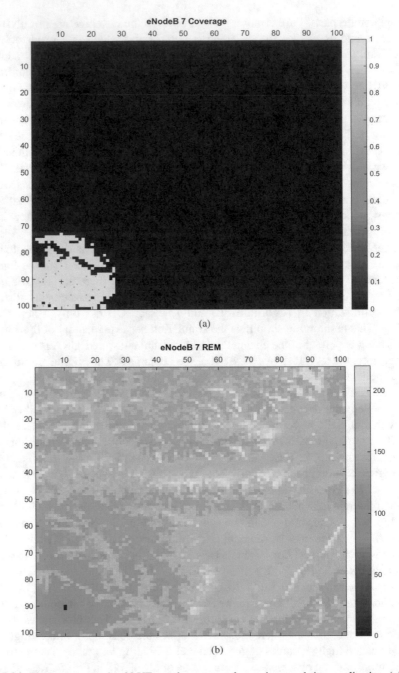

Fig. 7.24 The system contains 32 UEs, each concurrently running a real-time application. (**a**) eNB 7 Coverage. (**b**) eNB 7 REM

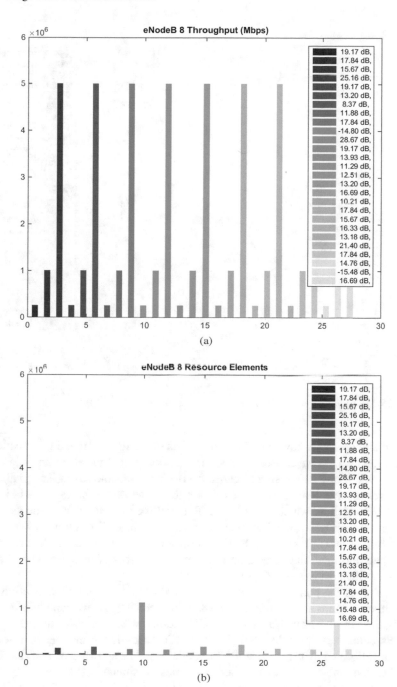

Fig. 7.25 The system contains 27 UEs, each concurrently running a real-time application. (**a**) Throughput of the UEs for eNB 8. (**b**) Resources Allocated to UEs by eNB 8

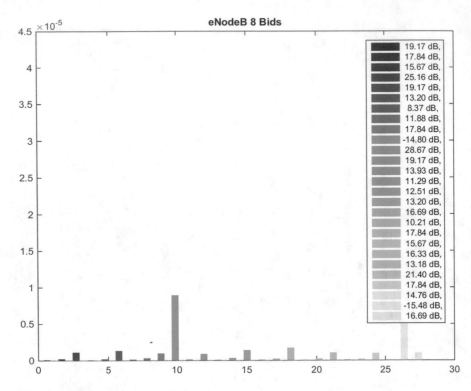

Fig. 7.26 UE Bids pledged to eNB 8

covered and are under coverage. The green dots in the region are the UEs scattered all over the area. As we can see, UEs are in the foot print of the eNBs and had good channel conditions. These simply chose eNB 8 as it was the strongest signal they could ever receive. Furthermore, we can see the REM which shows that a blue dot at the eNB location, which is for the eNB 8 and the blue color from the side bar represents a 0 dB loss which is expected as we are at the eNB position. However, as we go further away from the eNB, the path loss increases, which is shown by the color spectrum light blue, green, and yellow in the order of increase. The highest pathloss in the area is about 221 dB.

For eNB 9, the plots of resource elements allocated to the UEs as well as the throughput in Mbps are given in Fig. 7.28a,b, respectively. As we can observe from the figures, the horizontal axis is the UE indices which indicate that there are 28 UEs being served by the eNB 9. The vertical axis is the number of resource elements allocated by the optimization in Algorithms 12 and 13. Each UE has only 1 application running. This assumption simplifies the simulation because the goal of this chapter is observing the effect of the channel for which channel-aware EURA is performed. Furthermore, the legend shows the SNR of the UEs in dB. The bit rate requirements of the applications are according to $\{0.25, 1, 5, 0.25, 1, 5, \ldots\}$ Mbps.

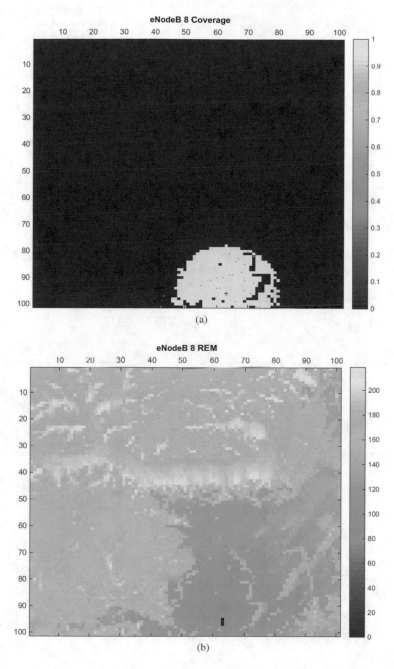

Fig. 7.27 The system contains 27 UEs, each concurrently running a real-time application. (**a**) eNB 8 Coverage. (**b**) eNB 8 REM

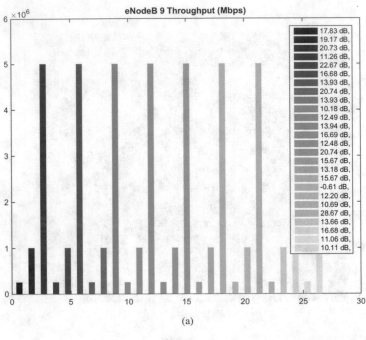

(a)

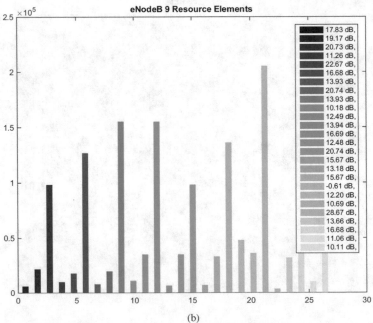

(b)

Fig. 7.28 The system contains 26 UEs, each concurrently running a real-time application. (**a**) Throughput of the UEs for eNB 9. (**b**) Resources Allocated to UEs by eNB 9

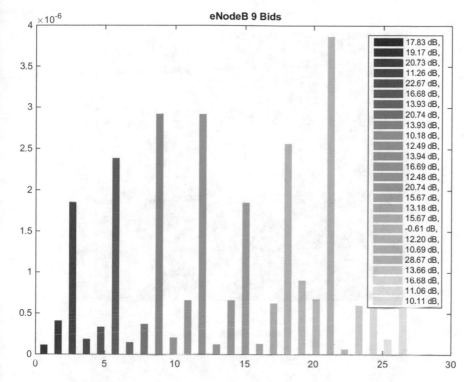

Fig. 7.29 UE Bids pledged to eNB 9

As we can observe from Fig. 7.28b, the UEs with low SNRs are receiving more resource elements in order to meet their bit rate requirements. On the other hand, UEs with high SNR are receiving less resources. On the other hand, Fig. 7.28a shows that UE throughputs are met. This is due all UEs are at good channel conditions such that the minimum SNR was -0.6 dB. We can see these in Fig. 7.28b that UEs with higher bit rate needs were allocated more resources, i.e. for UE k, $R_{1+3K} < R_{2+3K} < R_{3+3K}$. This statement says that the rate allocated to UE 1, 4, 7, ... is less than the rate allocated to UEs 2, 5, 8, ... and is less than the rates allocated to UEs 3, 6, 9, ... compared one to one (i.e. corresponding indices compared to each other). Also, we see that for same bit rate requirements, UEs with higher SNR are allocated less resources. For instance, UEs 3 and 15 are allocated less REs compared to UEs 6, 9, 12, 18, and 21 since they are at lower SNRs. In particular, the spike for UE 20 due to its low SNR at -0.6 dB.

Moreover, as we can see from Fig. 7.29, UEs $3 + 3K$ are higher than UEs $2 + 3K$ which bid higher than UEs $1 + 3K$ due to the fact that they require more resources in view of their applications (5 Mbps vs. 1 Mbps vs 0.25 Mbps). This plot shows the last iteration of the algorithm where the shadow prices are converged. Interestingly, UE 15 that needed high resources at bad channel conditions is bidding higher than its counterparts. Furthermore, we can observe the coverage area of eNB 9 in Fig. 7.30a,

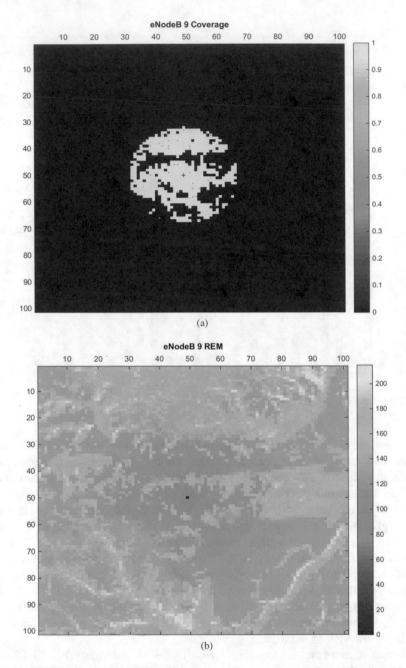

Fig. 7.30 The system contains 26 UEs, each concurrently running a real-time application. (**a**) eNB 9 Coverage. (**b**) eNB 9 REM

and as we see the coverage is not circular which is a result of REM and elevations in various directions leading to different SNRs. The black addition symbol in the middle of the coverage, shown in yellow, is the eNB 9 coordinates in the 101×101 grid. The side bar shows 0—dark blue—and 1—yellow—for the areas which are, respectively, not covered and are under coverage. The green dots in the region are the UEs scattered all over the area. As we can see, UEs are in the foot print of the eNBs and had good channel conditions. These simply chose eNB 9 as it was the strongest signal they could ever receive. Also, UE 15 is outside the footprint of the eNB which is reflected in its low SNR at -0.6 dB. Furthermore, we can see the REM which shows that a blue dot at the eNB location, which is for the eNB 9 and the blue color from the side bar represents a 0 dB loss which is expected as we are at the eNB position. However, as we go further away from the eNB, the path loss increases, which is shown by the color spectrum light blue, green, and yellow in the order of increase. The highest pathloss in the area is about 208 dB.

For eNB 10, the plots of resource elements allocated to the UEs as well as the throughput in Mbps are given in Fig. 7.31a,b, respectively. As we can observe from the figures, the horizontal axis is the UE indices which indicate that there are 14 UEs being served by the eNB 10. The vertical axis is the number of resource elements allocated by the optimization in Algorithms 12 and 13. Each UE has only 1 application running. This assumption simplifies the simulation because the goal of this chapter is observing the effect of the channel for which channel-aware EURA is performed. Furthermore, the legend shows the SNR of the UEs in dB. The bit rate requirements of the applications are according to $\{0.25, 1, 5, 0.25, 1, 5, \ldots\}$ Mbps.

As we can observe from Fig. 7.31b, the UEs with low SNRs are receiving more resource elements in order to meet their bit rate requirements. On the other hand, UEs with high SNR are receiving less resources. On the other hand, Fig. 7.31a shows that UE throughputs are met. This is due all UEs are at good channel conditions such that the minimum SNR was 5.62 dB. We can see these in Fig. 7.31b that UEs with higher bit rate needs were allocated more resources, i.e. for UE k, $R_{1+3K} < R_{2+3K} < R_{3+3K}$. This statement says that the rate allocated to UE 1, 4, 7, $\ldots$ is less than the rate allocated to UEs 2, 5, 8, $\ldots$ and is less than the rates allocated to UEs 3, 6, 9, $\ldots$ compared one to one (i.e. corresponding indices compared to each other). Also, we see that for same bit rate requirements, UEs with higher SNR are allocated less resources. For instance, UEs 3 and 12 are allocated more REs than UEs 6 and 9 since they are at lower SNRs. In particular, the spike for UE 14 is interesting as opposed to other rates of UEs R_{2+3K}. The spike is because this UE is at the lowest SNR situation in the system 5.62 dB.

Moreover, as we can see from Fig. 7.32, UEs $3 + 3K$ are higher than UEs $2 + 3K$ which bid higher than UEs $1 + 3K$ due to the fact that they require more resources in view of their applications (5 Mbps vs. 1 Mbps vs 0.25 Mbps). This plot shows the last iteration of the algorithm where the shadow prices are converged. Interestingly, UE 14 that needed high resources at good channel is bidding less than its counterparts. Furthermore, we can observe the coverage area of eNB 10 in Fig. 7.33a, and as we see the coverage is not circular which is a result of REM and elevations in various directions leading to different SNRs. The black addition

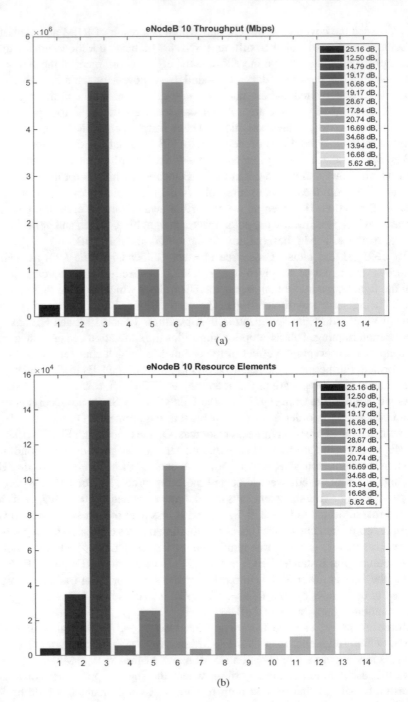

Fig. 7.31 The system contains 14 UEs, each concurrently running a real-time application. (**a**) Throughput of the UEs for eNB 10. (**b**) Resources Allocated to UEs by eNB 10

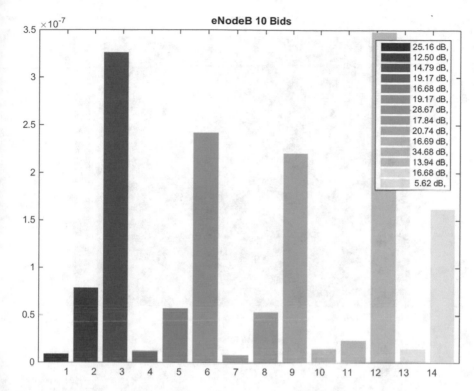

Fig. 7.32 UE Bids pledged to eNB 10

symbol in the middle of the coverage, shown in yellow, is the eNB 10 coordinates in the 101 × 101 grid. The side bar shows 0—dark blue—and 1—yellow—for the areas which are, respectively, not covered and are under coverage. The green dots in the region are the UEs scattered all over the area. As we can see, UEs are in the foot print of the eNBs and had good channel conditions. These simply chose eNB 10 as it was the strongest signal they could ever receive. Furthermore, we can see the REM which shows that a blue dot at the eNB location, which is for the eNB 10 and the blue color from the side bar represents a 0 dB loss which is expected as we are at the eNB position. We see that UE 14 which has the lowest SNR is outside the footprint of the eNB. However, as we go further away from the eNB, the path loss increases, which is shown by the color spectrum light blue, green, and yellow in the order of increase. The highest pathloss in the area is about 210 dB.

For eNB 11, the plots of resource elements allocated to the UEs as well as the throughput in Mbps are given in Fig. 7.34a,b, respectively. As we can observe from the figures, the horizontal axis is the UE indices which indicate that there are 24 UEs being served by the eNB 11. The vertical axis is the number of resource elements allocated by the optimization in Algorithms 12 and 13. Each UE has only 1 application running. This assumption simplifies the simulation because the goal of

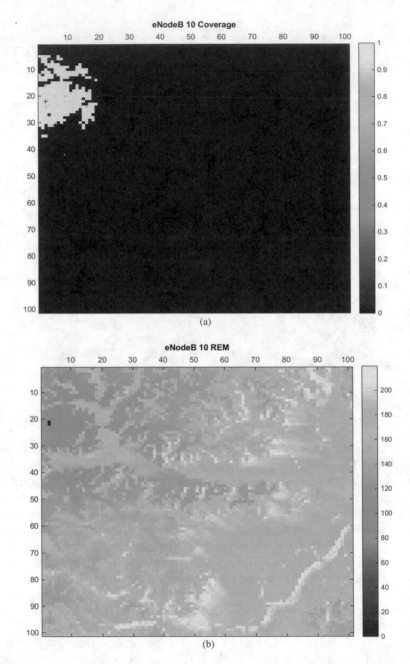

Fig. 7.33 The system contains 14 UEs, each concurrently running a real-time application. (**a**) eNB 10 Coverage. (**b**) eNB 10 REM

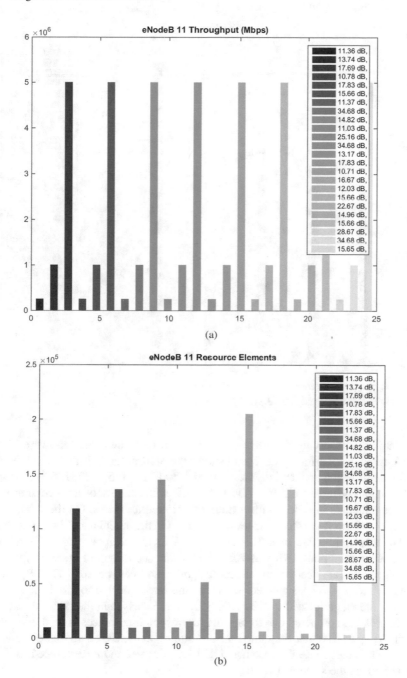

Fig. 7.34 The system contains 24 UEs, each concurrently running a real-time application. (**a**) Throughput of the UEs for eNB 11. (**b**) Resources Allocated to UEs by eNB 11

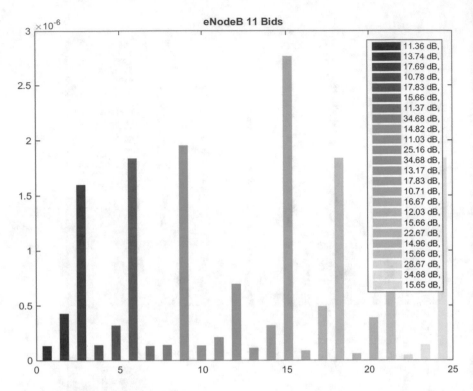

Fig. 7.35 UE Bids pledged to eNB 11

this chapter is observing the effect of the channel for which channel-aware EURA is performed. Furthermore, the legend shows the SNR of the UEs in dB. The bit rate requirements of the applications are according to {0.25, 1, 5, 0.25, 1, 5, . . .} Mbps.

As we can observe from Fig. 7.34b, the UEs with low SNRs are receiving more resource elements in order to meet their bit rate requirements. On the other hand, UEs with high SNR are receiving less resources. On the other hand, Fig. 7.34a shows that UE throughputs are met. This is due all UEs are at good channel conditions such that the minimum SNR was 10.7 dB. We can see these in Fig. 7.34b that UEs with higher bit rate needs were allocated more resources, i.e. for UE k, $R_{1+3K} <$ $R_{2+3K} < R_{3+3K}$. This statement says that the rate allocated to UE 1, 4, 7, . . . is less than the rate allocated to UEs 2, 5, 8, . . . and is less than the rates allocated to UEs 3, 6, 9, . . . compared one to one (i.e. corresponding indices compared to each other). Also, we see that for same bit rate requirements, UEs with higher SNR are allocated less resources. For instance, UE 15 is allocated more REs since it is at the lowest SNR in the system 10.7 dB.

Moreover, as we can see from Fig. 7.35, UEs $3 + 3K$ are higher than UEs $2 + 3K$ which bid higher than UEs $1 + 3K$ due to the fact that they require more resources in view of their applications (5 Mbps vs. 1 Mbps vs 0.25 Mbps). This plot shows the last iteration of the algorithm where the shadow prices are converged.

Interestingly, UE 15 that needed high resources at good channel is bidding less than its counterparts. Furthermore, we can observe the coverage area of eNB 11 in Fig. 7.36a, and as we see the coverage is not circular which is a result of REM and elevations in various directions leading to different SNRs. The black addition symbol in the middle of the coverage, shown in yellow, is the eNB 11 coordinates in the 101×101 grid. The side bar shows 0—dark blue—and 1—yellow—for the areas which are, respectively, not covered and are under coverage. The green dots in the region are the UEs scattered all over the area. As we can see, UEs are in the foot print of the eNBs and had good channel conditions. These simply chose eNB 11 as it was the strongest signal they could ever receive. Furthermore, we can see the REM which shows that a blue dot at the eNB location, which is for the eNB 11 and the blue color from the side bar represents a 0 dB loss which is expected as we are at the eNB position. However, as we go further away from the eNB, the path loss increases, which is shown by the color spectrum light blue, green, and yellow in the order of increase. The highest pathloss in the area is about 206 dB.

For eNB 12, the plots of resource elements allocated to the UEs as well as the throughput in Mbps are given in Fig. 7.37a,b, respectively. As we can observe from the figures, the horizontal axis is the UE indices which indicate that there are 25 UEs being served by the eNB 12. The vertical axis is the number of resource elements allocated by the optimization in Algorithms 12 and 13. Each UE has only 1 application running. This assumption simplifies the simulation because the goal of this chapter is observing the effect of the channel for which channel-aware EURA is performed. Furthermore, the legend shows the SNR of the UEs in dB. The bit rate requirements of the applications are according to $\{0.25, 1, 5, 0.25, 1, 5, \ldots\}$ Mbps.

As we can observe from Fig. 7.37b, the UEs with low SNRs are receiving more resource elements in order to meet their bit rate requirements. On the other hand, UEs with high SNR are receiving less resources. On the other hand, Fig. 7.37a shows that UE throughputs are met. This is due all UEs are at good channel conditions such that the minimum SNR was -8.67 dB. We can see these in Fig. 7.37b that UEs with higher bit rate needs were allocated more resources, i.e. for UE k, $R_{1+3K} < R_{2+3K} < R_{3+3K}$. This statement says that the rate allocated to UE 1, 4, 7, ... is less than the rate allocated to UEs 2, 5, 8, ... and is less than the rates allocated to UEs 3, 6, 9, ... compared one to one (i.e. corresponding indices compared to each other). Also, we see that for same bit rate requirements, UEs with higher SNR are allocated less resources. For instance, UEs 11, 12, and 18 are allocated more REs since they are at lower SNRs. In particular, the spike for UE 8 is interesting since it is at the lowest SNR situation in the system -8.67 dB.

Moreover, as we can see from Fig. 7.38, UEs $3 + 3K$ are higher than UEs $2 + 3K$ which bid higher than UEs $1 + 3K$ due to the fact that they require more resources in view of their applications (5 Mbps vs. 1 Mbps vs 0.25 Mbps). This plot shows the last iteration of the algorithm where the shadow prices are converged. Interestingly, UE 8 that needed high resources at bad channel is bidding higher than its counterparts. Furthermore, we can observe the coverage area of eNB 12 in Fig. 7.39a, and as we see the coverage is not circular which is a result of REM and elevations in various directions leading to different SNRs. The black addition

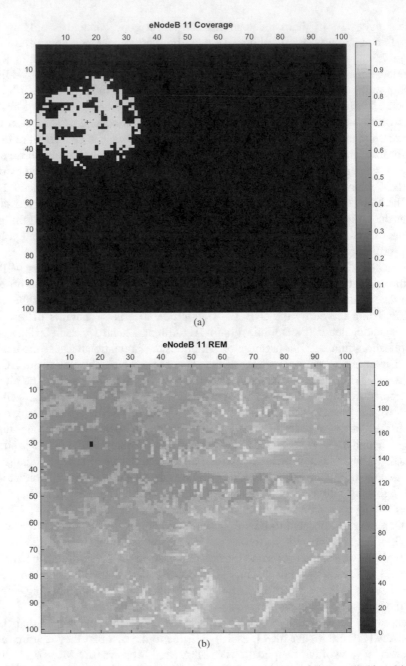

Fig. 7.36 The system contains 24 UEs, each concurrently running a real-time application. (**a**) eNB 11 Coverage. (**b**) eNB 11 REM

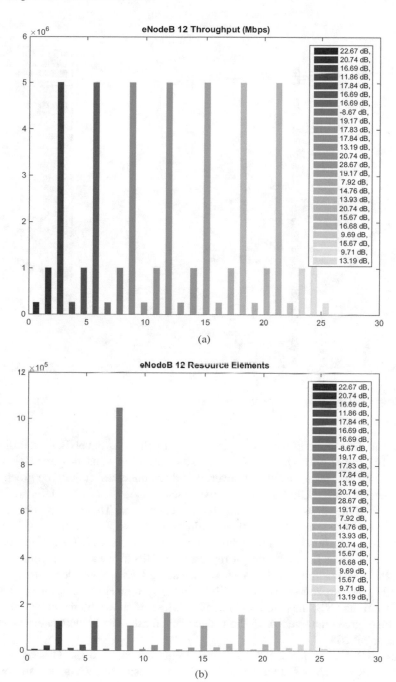

Fig. 7.37 The system contains 25 UEs, each concurrently running a real-time application. (**a**) Throughput of the UEs for eNB 12. (**b**) Resources Allocated to UEs by eNB 12

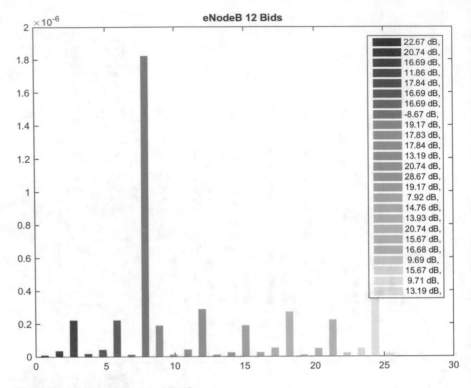

Fig. 7.38 UE Bids pledged to eNB 12

symbol in the middle of the coverage, shown in yellow, is the eNB 12 coordinates in the 101×101 grid. The side bar shows 0—dark blue—and 1—yellow—for the areas which are, respectively, not covered and are under coverage. The green dots in the region are the UEs scattered all over the area. As we can see, UEs are in the foot print of the eNBs and had good channel conditions. These simply chose eNB 12 as it was the strongest signal they could ever receive. Furthermore, we can see the REM which shows that a blue dot at the eNB location, which is for the eNB 12 and the blue color from the side bar represents a 0 dB loss which is expected as we are at the eNB position. Furthermore, we see some UEs are outside the footprint of the eNB which are reflected in lower SNRs in the system. However, as we go further away from the eNB, the path loss increases, which is shown by the color spectrum light blue, green, and yellow in the order of increase. The highest pathloss in the area is about 205 dB.

For eNB 13, the plots of resource elements allocated to the UEs as well as the throughput in Mbps are given in Fig. 7.40a,b, respectively. As we can observe from the figures, the horizontal axis is the UE indices which indicate that there are 21 UEs being served by the eNB 13. The vertical axis is the number of resource elements allocated by the optimization in Algorithms 12 and 13. Each UE has only 1 application running. This assumption simplifies the simulation because the goal of

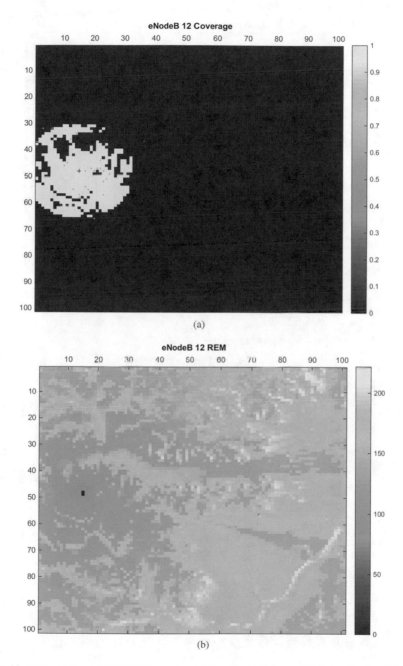

Fig. 7.39 The system contains 25 UEs, each concurrently running a real-time application. (**a**) cNB 12 Coverage. (**b**) eNB 12 REM

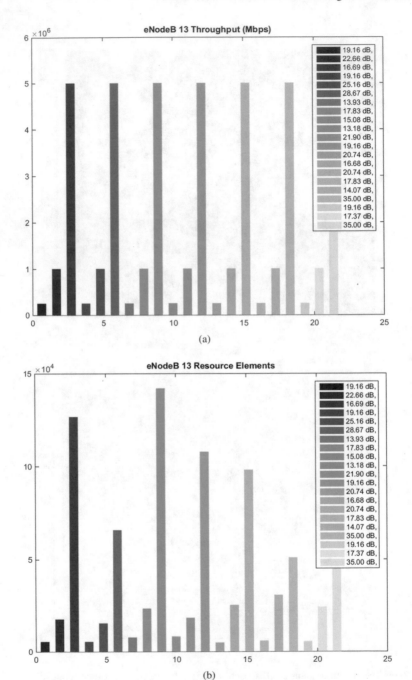

Fig. 7.40 The system contains 21 UEs, each concurrently running a real-time application. (**a**) Throughput of the UEs for eNB 13. (**b**) Resources Allocated to UEs by eNB 13

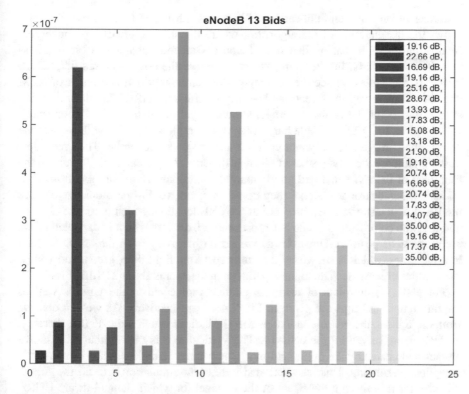

Fig. 7.41 UE Bids pledged to eNB 13

this chapter is observing the effect of the channel for which channel-aware EURA is performed. Furthermore, the legend shows the SNR of the UEs in dB. The bit rate requirements of the applications are according to $\{0.25, 1, 5, 0.25, 1, 5, \ldots\}$ Mbps.

As we can observe from Fig. 7.40b, the UEs with low SNRs are receiving more resource elements in order to meet their bit rate requirements. On the other hand, UEs with high SNR are receiving less resources. On the other hand, Fig. 7.40a shows that UE throughputs are met. This is due all UEs are at good channel conditions such that the minimum SNR was 13.18 dB. We can see these in Fig. 7.40b that UEs with higher bit rate needs were allocated more resources, i.e. for UE k, $R_{1+3K} < R_{2+3K} < R_{3+3K}$. This statement says that the rate allocated to UE 1, 4, 7, ... is less than the rate allocated to UEs 2, 5, 8, ... and is less than the rates allocated to UEs 3, 6, 9, ... compared one to one (i.e. corresponding indices compared to each other). Also, we see that for same bit rate requirements, UEs with higher SNR are allocated less resources. For instance, UE 9 is allocated more REs since it is at lower SNRs. In particular, the spike for UE 9 is interesting because this UE is at the lowest SNR situation in the system 13.18 dB.

Moreover, as we can see from Fig. 7.41, UEs $3 + 3K$ are higher than UEs $2 + 3K$ which bid higher than UEs $1 + 3K$ due to the fact that they require more resources in view of their applications (5 Mbps vs. 1 Mbps vs 0.25 Mbps). This plot shows the last iteration of the algorithm where the shadow prices are converged. Interestingly, UE 9 that needed high resources at good channel is bidding less than its counterparts. Furthermore, we can observe the coverage area of eNB 13 in Fig. 7.15a, and as we see the coverage is not circular which is a result of REM and elevations in various directions leading to different SNRs. The black addition symbol in the middle of the coverage, shown in yellow, is the eNB 4 coordinates in the 101×101 grid. The side bar shows 0—dark blue—and 1—yellow—for the areas which are, respectively, not covered and are under coverage. The green dots in the region are the UEs scattered all over the area. As we can see, UEs are in the foot print of the eNBs and had good channel conditions. These simply chose eNB 13 as it was the strongest signal they could ever receive. Furthermore, we can see the REM which shows that a blue dot at the eNB location, which is for the eNB 13 and the blue color from the side bar represents a 0 dB loss which is expected as we are at the eNB position. However, as we go further away from the eNB, the path loss increases, which is shown by the color spectrum light blue, green, and yellow in the order of increase. The highest pathloss in the area is about 218 dB (Fig. 7.42).

For eNB 14, the plots of resource elements allocated to the UEs as well as the throughput in Mbps are given in Fig. 7.43a,b, respectively. As we can observe from the figures, the horizontal axis is the UE indices which indicate that there are 25 UEs being served by the eNB 14. The vertical axis is the number of resource elements allocated by the optimization in Algorithms 12 and 13. Each UE has only 1 application running. This assumption simplifies the simulation because the goal of this chapter is observing the effect of the channel for which channel-aware EURA is performed. Furthermore, the legend shows the SNR of the UEs in dB. The bit rate requirements of the applications are according to $\{0.25, 1, 5, 0.25, 1, 5, \ldots\}$ Mbps.

As we can observe from Fig. 7.43b, the UEs with low SNRs are receiving more resource elements in order to meet their bit rate requirements. On the other hand, UEs with high SNR are receiving less resources. On the other hand, Fig. 7.43a shows that UE throughputs are met. This is due all UEs are at good channel conditions such that the minimum SNR was 11.25 dB. We can see these in Fig. 7.43b that UEs with higher bit rate needs were allocated more resources, i.e. for UE k, $R_{1+3K} < R_{2+3K} < R_{3+3K}$. This statement says that the rate allocated to UE 1, 4, 7, ... is less than the rate allocated to UEs 2, 5, 8, ... and is less than the rates allocated to UEs 3, 6, 9, ... compared one to one (i.e. corresponding indices compared to each other). Also, we see that for same bit rate requirements, UEs with higher SNR are allocated less resources. For instance, UE 18 who has a high bit requirement of 5 Mbps is getting more REs than its counterparts since it is at a bad channel condition 11.25 dB.

Moreover, as we can see from Fig. 7.44, UEs $3 + 3K$ are higher than UEs $2 + 3K$ which bid higher than UEs $1 + 3K$ due to the fact that they require more resources in view of their applications (5 Mbps vs. 1 Mbps vs 0.25 Mbps). This plot shows the last iteration of the algorithm where the shadow prices are converged. Interestingly, UE 18 that needed high resources at good channel is bidding higher than its counterparts since it is at a bad channel condition. Furthermore, we can

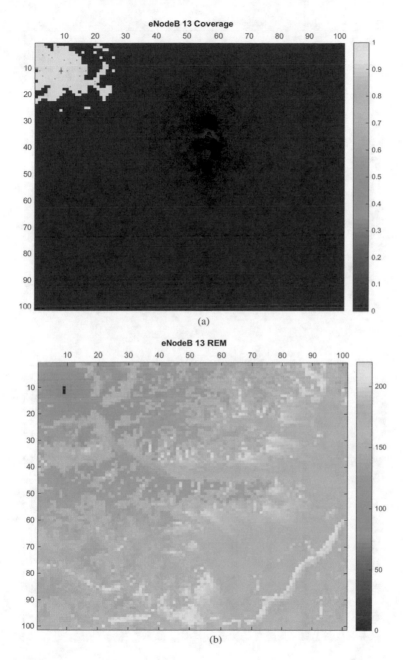

Fig. 7.42 The system contains 21 UEs, each concurrently running a real-time application. (**a**) eNB 13 Coverage. (**b**) eNB 13 REM

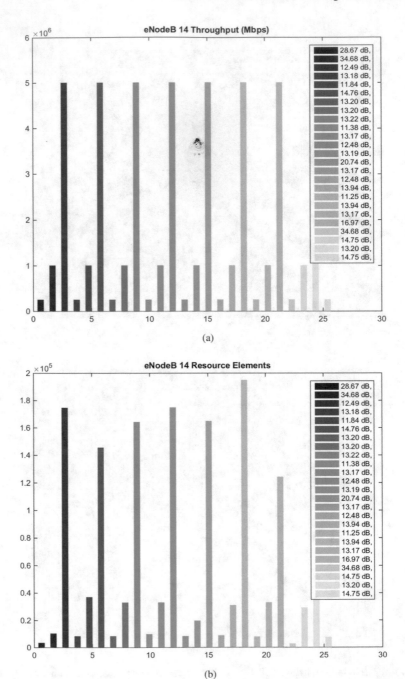

Fig. 7.43 The system contains 25 UEs, each concurrently running a real-time application. (**a**) Throughput of the UEs for eNB 14. (**b**) Resources Allocated to UEs by eNB 14

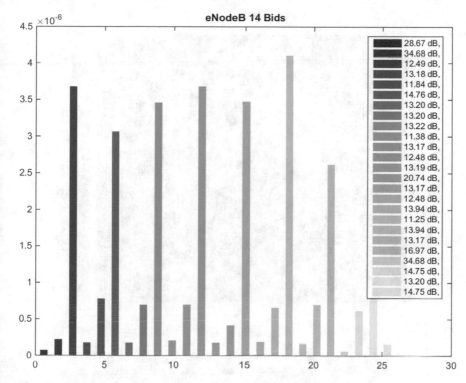

Fig. 7.44 UE Bids pledged to eNB 14

observe the coverage area of eNB 14 in Fig. 7.45a, and as we see the coverage is not circular which is a result of REM and elevations in various directions leading to different SNRs. The black addition symbol in the middle of the coverage, shown in yellow, is the eNB 14 coordinates in the 101×101 grid. The side bar shows 0—dark blue—and 1—yellow—for the areas which are, respectively, not covered and are under coverage. The green dots in the region are the UEs scattered all over the area. As we can see, UEs are in the foot print of the eNBs and had good channel conditions. These simply chose eNB 14 as it was the strongest signal they could ever receive. Furthermore, we can see the REM which shows that a blue dot at the eNB location, which is for the eNB 14 and the blue color from the side bar represents a 0 dB loss which is expected as we are at the eNB position. However, as we go further away from the eNB, the path loss increases, which is shown by the color spectrum light blue, green, and yellow in the order of increase. The highest pathloss in the area is about 209 dB.

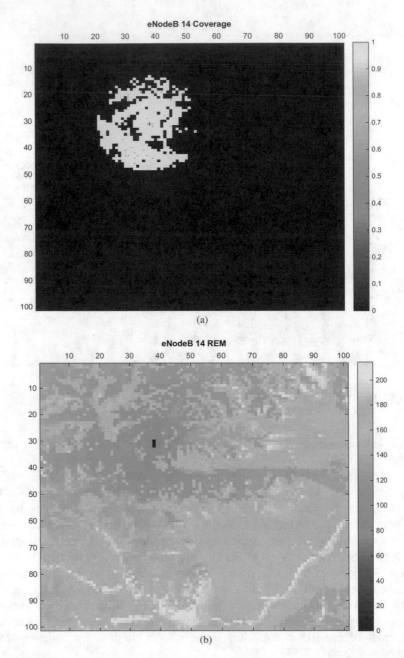

Fig. 7.45 The system contains 25 UEs, each concurrently running a real-time application. (**a**) eNB 14 Coverage. (**b**) eNB 14 REM

For eNB 15, the plots of resource elements allocated to the UEs as well as the throughput in Mbps are given in Fig. 7.46a,b, respectively. As we can observe from the figures, the horizontal axis is the UE indices which indicate that there are 31 UEs being served by the eNB 15. The vertical axis is the number of resource elements allocated by the optimization in Algorithms 12 and 13. Each UE has only 1 application running. This assumption simplifies the simulation because the goal of this chapter is observing the effect of the channel for which channel-aware EURA is performed. Furthermore, the legend shows the SNR of the UEs in dB. The bit rate requirements of the applications are according to $\{0.25, 1, 5, 0.25, 1, 5, \ldots\}$ Mbps.

As we can observe from Fig. 7.46b, the UEs with low SNRs are receiving more resource elements in order to meet their bit rate requirements. On the other hand, UEs with high SNR are receiving less resources. On the other hand, Fig. 7.46a shows that UE throughputs are met. This is due all UEs are at good channel conditions such that the minimum SNR was -2.14 dB. We can see these in Fig. 7.46b that UEs with higher bit rate needs were allocated more resources, i.e. for UE k, $R_{1+3K} < R_{2+3K} < R_{3+3K}$. This statement says that the rate allocated to UE 1, 4, 7, $\ldots$ is less than the rate allocated to UEs 2, 5, 8, $\ldots$ and is less than the rates allocated to UEs 3, 6, 9, $\ldots$ compared one to one (i.e. corresponding indices compared to each other). Also, we see that for same bit rate requirements, UEs with higher SNR are allocated less resources. For instance, UEs 12 and 24 are allocated more REs since they are at lower SNRs. In particular, the spike for UE 10 is interesting even though it has less bit needs (0.25 Mbps). The spike is because this UE is at the lowest SNR situation in the system -2.14 dB.

Moreover, as we can see from Fig. 7.47, UEs $3 + 3K$ are higher than UEs $2 + 3K$ which bid higher than UEs $1 + 3K$ due to the fact that they require more resources in view of their applications (5 Mbps vs. 1 Mbps vs 0.25 Mbps). This plot shows the last iteration of the algorithm where the shadow prices are converged. Interestingly, UE 15 that needed high resources at good channel is bidding less than its counterparts. Furthermore, we can observe the coverage area of eNB 15 in Fig. 7.48a, and as we see the coverage is not circular which is a result of REM and elevations in various directions leading to different SNRs. The black addition symbol in the middle of the coverage, shown in yellow, is the eNB 15 coordinates in the 101×101 grid. The side bar shows 0—dark blue—and 1—yellow—for the areas which are, respectively, not covered and are under coverage. The green dots in the region are the UEs scattered all over the area. As we can see, UEs are not just in the foot print of the eNBs, and these represent the lowest SNRs in the system. These simply chose eNB 15 as it was the strongest signal they could ever receive. Furthermore, we can see the REM which shows that a blue dot at the eNB location, which is for the eNB 15 and the blue color from the side bar represents a 0 dB loss which is expected as we are at the eNB position. However, as we go further away from the eNB, the path loss increases, which is shown by the color spectrum light blue, green, and yellow in the order of increase. The highest pathloss in the area is about 209 dB.

For eNB 16, the plots of resource elements allocated to the UEs as well as the throughput in Mbps are given in Fig. 7.49a,b, respectively. As we can observe

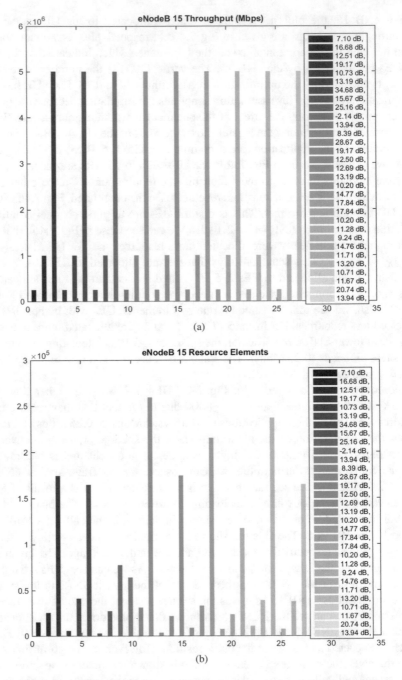

Fig. 7.46 The system contains 31 UEs, each concurrently running a real-time application. (**a**) Throughput of the UEs for eNB 15. (**b**) Resources Allocated to UEs by eNB 15

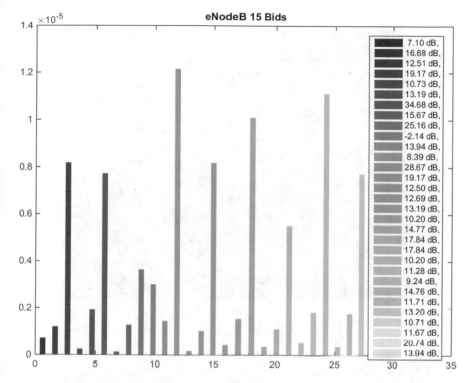

Fig. 7.47 UE Bids pledged to eNB 15

from the figures, the horizontal axis is the UE indices which indicate that there are 14 UEs being served by the eNB 16. The vertical axis is the number of resource elements allocated by the optimization in Algorithms 12 and 13. Each UE has only 1 application running. This assumption simplifies the simulation because the goal of this chapter is observing the effect of the channel for which channel-aware EURA is performed. Furthermore, the legend shows the SNR of the UEs in dB. The bit rate requirements of the applications are according to $\{0.25, 1, 5, 0.25, 1, 5, \ldots\}$ Mbps.

As we can observe from Fig. 7.49b, the UEs with low SNRs are receiving more resource elements in order to meet their bit rate requirements. On the other hand, UEs with high SNR are receiving less resources. On the other hand, Fig. 7.49a shows that UE throughputs are met. This is due all UEs are at good channel conditions such that the minimum SNR was $-12.30\,$dB. We can see these in Fig. 7.19b that UEs with higher bit rate needs were allocated more resources, i.e. for UE k, $R_{1+3K} < R_{2+3K} < R_{3+3K}$. This statement says that the rate allocated to UE 1, 4, 7, ... is less than the rate allocated to UEs 2, 5, 8, ... and is less than the rates allocated to UEs 3, 6, 9, ... compared one to one (i.e. corresponding indices compared to each other). Also, we see that for same bit rate requirements, UEs with higher SNR are allocated less resources. For instance, UE 4 is allocated more REs since it is at the lowest SNR $-12.30\,$dB.

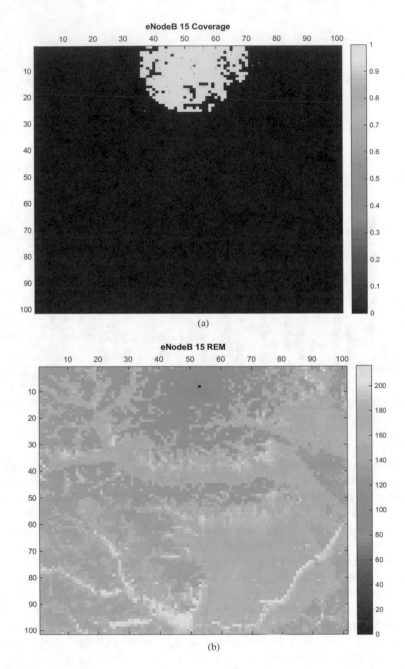

Fig. 7.48 The system contains 31 UEs, each concurrently running a real-time application. (**a**) eNB 15 Coverage. (**b**) eNB 15 REM

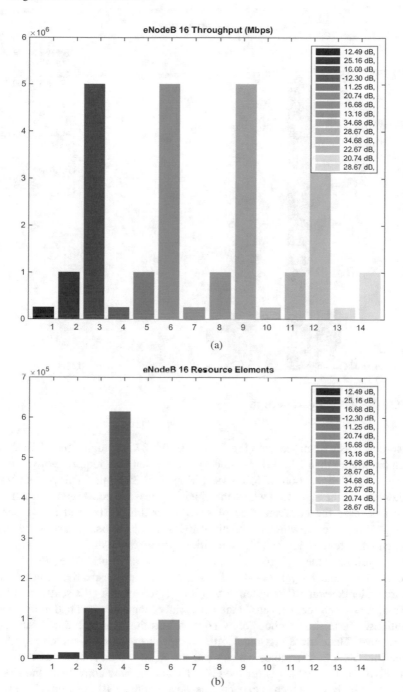

Fig. 7.49 The system contains 14 UEs, each concurrently running a real-time application. (**a**) Throughput of the UEs for eNB 16. (**b**) Resources Allocated to UEs by eNB 16

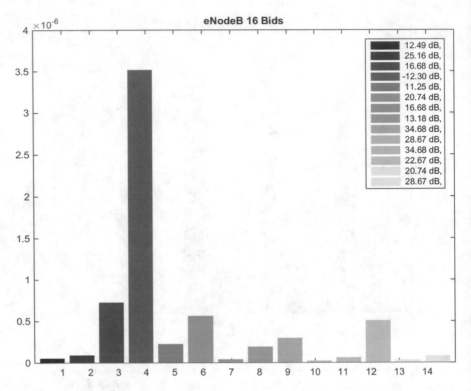

Fig. 7.50 UE Bids pledged to eNB 16

Moreover, as we can see from Fig. 7.50, UEs $3+3K$ are higher than UEs $2+3K$ which bid higher than UEs $1+3K$ due to the fact that they require more resources in view of their applications (5 Mbps vs. 1 Mbps vs 0.25 Mbps). This plot shows the last iteration of the algorithm where the shadow prices are converged. Furthermore, we can observe the coverage area of eNB 16 in Fig. 7.51a, and as we see the coverage is not circular which is a result of REM and elevations in various directions leading to different SNRs. The black addition symbol in the middle of the coverage, shown in yellow, is the eNB 16 coordinates in the 101×101 grid. The side bar shows 0—dark blue—and 1—yellow—for the areas which are, respectively, not covered and are under coverage. The green dots in the region are the UEs scattered all over the area. As we can see, UEs are in the foot print of the eNBs and had good channel conditions. These simply chose eNB 16 as it was the strongest signal they could ever receive. There are SNRs outside the footprint of the eNB, which is reflected by the low SNR Furthermore, we can see the REM which shows that a blue dot at the eNB location, which is for the eNB 16 and the blue color from the side bar represents a 0 dB loss which is expected as we are at the eNB position. However, as we go further away from the eNB, the path loss increases, which is shown by the

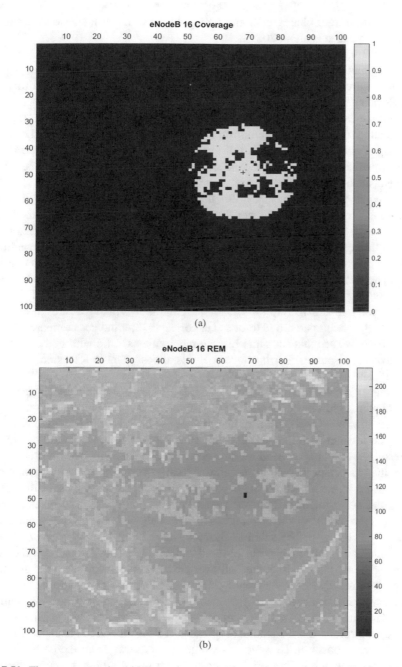

Fig. 7.51 The system contains 14 UEs, each concurrently running a real-time application. (**a**) eNB 16 Coverage. (**b**) eNB 16 REM

color spectrum light blue, green, and yellow in the order of increase. The highest pathloss in the area is about 205 dB.

For eNB 17, the plots of resource elements allocated to the UEs as well as the throughput in Mbps are given in Fig. 7.52a,b, respectively. As we can observe from the figures, the horizontal axis is the UE indices which indicate that there are 18 UEs being served by the eNB 18. The vertical axis is the number of resource elements allocated by the optimization in Algorithms 12 and 13. Each UE has only 1 application running. This assumption simplifies the simulation because the goal of this chapter is observing the effect of the channel for which channel-aware EURA is performed. Furthermore, the legend shows the SNR of the UEs in dB. The bit rate requirements of the applications are according to $\{0.25, 1, 5, 0.25, 1, 5, \ldots\}$ Mbps.

As we can observe from Fig. 7.52b, the UEs with low SNRs are receiving more resource elements in order to meet their bit rate requirements. On the other hand, UEs with high SNR are receiving less resources. On the other hand, Fig. 7.52a shows that UE throughputs are met. This is due all UEs are at good channel conditions such that the minimum SNR was 8.68 dB. We can see these in Fig. 7.52b that UEs with higher bit rate needs were allocated more resources, i.e. for UE k, $R_{1+3K} < R_{2+3K} < R_{3+3K}$. This statement says that the rate allocated to UE 1, 4, 7, ... is less than the rate allocated to UEs 2, 5, 8, ... and is less than the rates allocated to UEs 3, 6, 9, ... compared one to one (i.e. corresponding indices compared to each other). Also, we see that for same bit rate requirements, UEs with higher SNR are allocated less resources. For instance, UE 9 is allocated more REs since than UEs 3, 12, 15, and 18 as it is at lower SNRs.

Moreover, as we can see from Fig. 7.53, UEs $3 + 3K$ are higher than UEs $2 + 3K$ which bid higher than UEs $1 + 3K$ due to the fact that they require more resources in view of their applications (5 Mbps vs. 1 Mbps vs 0.25 Mbps). This plot shows the last iteration of the algorithm where the shadow prices are converged. Interestingly, UE 9 that needed high resources at bad channel conditions is bidding higher than its counterparts. Furthermore, we can observe the coverage area of eNB 17 in Fig. 7.54a, and as we see the coverage is not circular which is a result of REM and elevations in various directions leading to different SNRs. The black addition symbol in the middle of the coverage, shown in yellow, is the eNB 17 coordinates in the 101×101 grid. The side bar shows 0—dark blue—and 1—yellow—for the areas which are, respectively, not covered and are under coverage. The green dots in the region are the UEs scattered all over the area. As we can see, UEs are in the foot print of the eNBs and had good channel conditions. These simply chose eNB 17 as it was the strongest signal they could ever receive. Furthermore, we can see the REM which shows that a blue dot at the eNB location, which is for the eNB 17 and the blue color from the side bar represents a 0 dB loss which is expected as we are at the eNB position. However, as we go further away from the eNB, the path loss increases, which is shown by the color spectrum light blue, green, and yellow in the order of increase. The highest pathloss in the area is about 210 dB.

For eNB 18, the plots of resource elements allocated to the UEs as well as the throughput in Mbps are given in Fig. 7.55a,b, respectively. As we can observe from the figures, the horizontal axis is the UE indices which indicate that there are

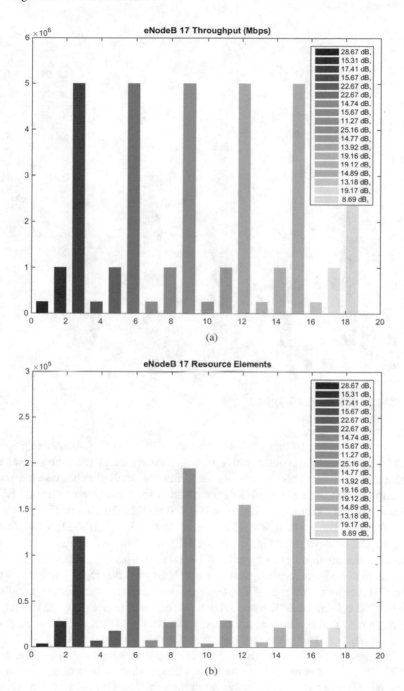

Fig. 7.52 The system contains 18 UEs, each concurrently running a real-time application. (**a**) Throughput of the UEs for eNB 17. (**b**) Resources Allocated to UEs by eNB 17

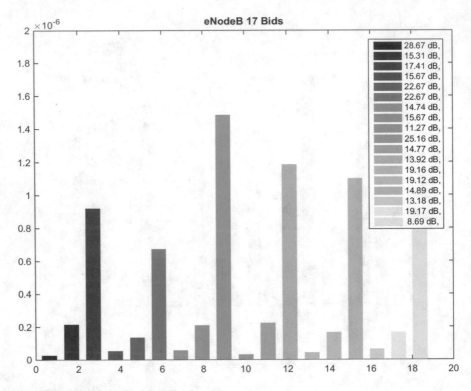

Fig. 7.53 UE Bids pledged to eNB 17

30 UEs being served by the eNB 4. The vertical axis is the number of resource elements allocated by the optimization in Algorithms 12 and 13. Each UE has only 1 application running. This assumption simplifies the simulation because the goal of this chapter is observing the effect of the channel for which channel-aware EURA is performed. Furthermore, the legend shows the SNR of the UEs in dB. The bit rate requirements of the applications are according to $\{0.25, 1, 5, 0.25, 1, 5, \ldots\}$ Mbps.

As we can observe from Fig. 7.55b, the UEs with low SNRs are receiving more resource elements in order to meet their bit rate requirements. On the other hand, UEs with high SNR are receiving less resources. On the other hand, Fig. 7.55a shows that UE throughputs are met. This is due all UEs are at good channel conditions such that the minimum SNR was 5.64 dB. We can see these in Fig. 7.55b that UEs with higher bit rate needs were allocated more resources, i.e. for UE k, $R_{1+3K} < R_{2+3K} < R_{3+3K}$. This statement says that the rate allocated to UE 1, 4, 7, … is less than the rate allocated to UEs 2, 5, 8, … and is less than the rates allocated to UEs 3, 6, 9, … compared one to one (i.e. corresponding indices compared to each other). Also, we see that for same bit rate requirements, UEs with higher SNR are allocated less resources. For instance, UEs 12, 21, 24, and 18 are allocated more REs since they are at lower SNRs with respect to their counterparts.

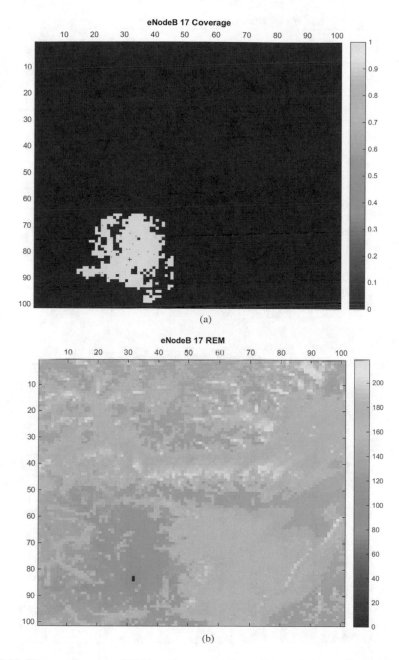

Fig. 7.54 The system contains 18 UEs, each concurrently running a real-time application. (**a**) eNB 17 Coverage. (**b**) eNB 17 REM

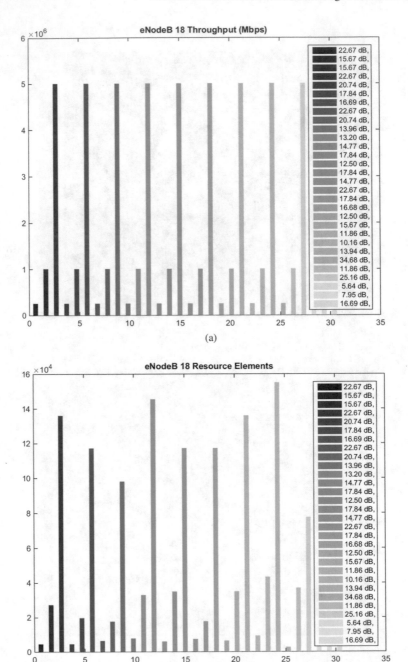

Fig. 7.55 The system contains 30 UEs, each concurrently running a real-time application. (**a**) Throughput of the UEs for eNB 18. (**b**) Resources Allocated to UEs by eNB 18

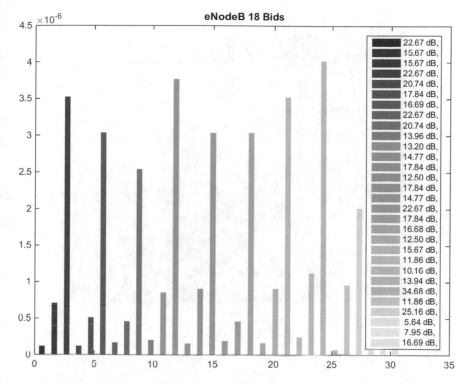

Fig. 7.56 UE Bids pledged to eNB 18

Moreover, as we can see from Fig. 7.56, UEs $3+3K$ are higher than UEs $2+3K$ which bid higher than UEs $1+3K$ due to the fact that they require more resources in view of their applications (5 Mbps vs. 1 Mbps vs 0.25 Mbps). This plot shows the last iteration of the algorithm where the shadow prices are converged. Interestingly, UE 24 that needed high resources at bad channel conditions is bidding higher than its counterparts. Furthermore, we can observe the coverage area of eNB 18 in Fig. 7.57a, and as we see the coverage is not circular which is a result of REM and elevations in various directions leading to different SNRs. The black addition symbol in the middle of the coverage, shown in yellow, is the eNB 18 coordinates in the 101×101 grid. The side bar shows 0—dark blue—and 1—yellow—for the areas which are, respectively, not covered and are under coverage. The green dots in the region are the UEs scattered all over the area. As we can see, UEs are in the foot print of the eNBs and had good channel conditions. These simply chose eNB 18 as it was the strongest signal they could ever receive. Furthermore, we can see the REM which shows that a blue dot at the eNB location, which is for the eNB 3 and the blue color from the side bar represents a 0 dB loss which is expected as we are at the eNB position. However, as we go further away from the eNB, the path loss increases, which is shown by the color spectrum light blue, green, and yellow in the order of increase. The highest pathloss in the area is about 217 dB.

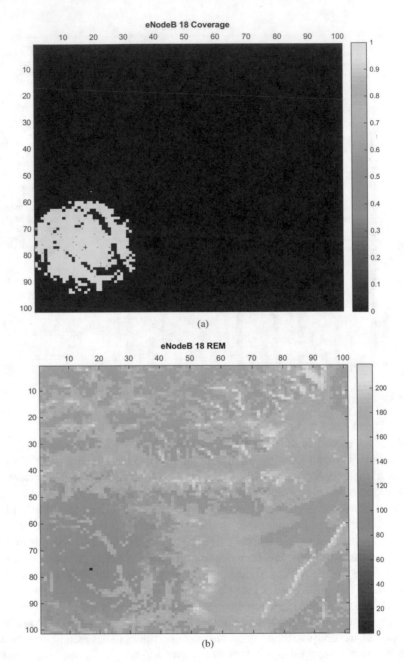

Fig. 7.57 The system contains 30 UEs, each concurrently running a real-time application. (**a**) eNB 18 Coverage. (**b**) eNB 18 REM

For eNB 19, the plots of resource elements allocated to the UEs as well as the throughput in Mbps are given in Fig. 7.58a,b, respectively. As we can observe from the figures, the horizontal axis is the UE indices which indicate that there are 23 UEs being served by the eNB 19. The vertical axis is the number of resource elements allocated by the optimization in Algorithms 12 and 13. Each UE has only 1 application running. This assumption simplifies the simulation because the goal of this chapter is observing the effect of the channel for which channel-aware EURA is performed. Furthermore, the legend shows the SNR of the UEs in dB. The bit rate requirements of the applications are according to $\{0.25, 1, 5, 0.25, 1, 5, \ldots\}$ Mbps.

As we can observe from Fig. 7.58b, the UEs with low SNRs are receiving more resource elements in order to meet their bit rate requirements. On the other hand, UEs with high SNR are receiving less resources. On the other hand, Fig. 7.58a shows that UE throughputs are met. This is due all UEs are at good channel conditions such that the minimum SNR was -18.95 dB. We can see these in Fig. 7.58b that UEs with higher bit rate needs were allocated more resources, i.e. for UE k, $R_{1+3K} < R_{2+3K} < R_{3+3K}$. This statement says that the rate allocated to UE 1, 4, 7, $\ldots$ is less than the rate allocated to UEs 2, 5, 8, $\ldots$ and is less than the rates allocated to UEs 3, 6, 9, $\ldots$ compared one to one (i.e. corresponding indices compared to each other). Also, we see that for same bit rate requirements, UEs with higher SNR are allocated less resources.

Moreover, as we can see from Fig. 7.59, UEs $3 + 3K$ are higher than UEs $2 + 3K$ which bid higher than UEs $1 + 3K$ due to the fact that they require more resources in view of their applications (5 Mbps vs. 1 Mbps vs 0.25 Mbps). This plot shows the last iteration of the algorithm where the shadow prices are converged. Interestingly, UE 15 that needed high resources at good channel is bidding less than its counterparts. Furthermore, we can observe the coverage area of eNB 19 in Fig. 7.60a, and as we see the coverage is not circular which is a result of REM and elevations in various directions leading to different SNRs. The black addition symbol in the middle of the coverage, shown in yellow, is the eNB 19 coordinates in the 101×101 grid. The side bar shows 0—dark blue—and 1—yellow—for the areas which are, respectively, not covered and are under coverage. The green dots in the region are the UEs scattered all over the area.

For eNB 20, the plots of resource elements allocated to the UEs as well as the throughput in Mbps are given in Fig. 7.61a,b, respectively. As we can observe from the figures, the horizontal axis is the UE indices which indicate that there are 32 UEs being served by the eNB 4. The vertical axis is the number of resource elements allocated by the optimization in Algorithms 12 and 13. Each UE has only 1 application running. This assumption simplifies the simulation because the goal of this chapter is observing the effect of the channel for which channel-aware EURA is performed. Furthermore, the legend shows the SNR of the UEs in dB. The bit rate requirements of the applications are according to $\{0.25, 1, 5, 0.25, 1, 5, \ldots\}$ Mbps.

As we can observe from Fig. 7.61b, the UEs with low SNRs are receiving more resource elements in order to meet their bit rate requirements. On the other hand, UEs with high SNR are receiving less resources. On the other hand, Fig. 7.61a shows that UE throughputs are met. This is due all UEs are at good channel conditions

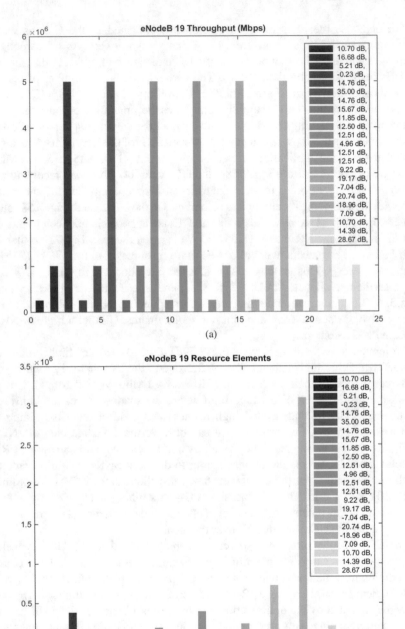

Fig. 7.58 The system contains 23 UEs, each concurrently running a real-time application. (**a**) Throughput of the UEs for eNB 19. (**b**) Resources Allocated to UEs by eNB 19

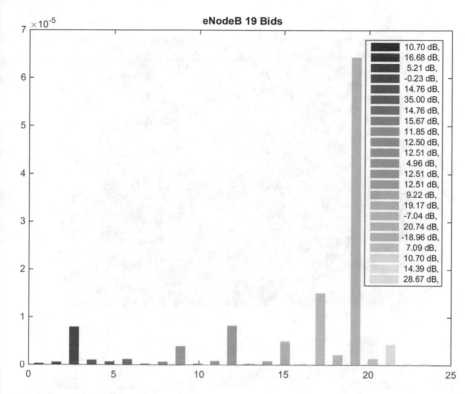

Fig. 7.59 UE Bids pledged to eNB 19

such that the minimum SNR was 12.5 dB. We can see these in Fig. 7.13b that UEs with higher bit rate needs were allocated more resources, i.e. for UE k, $R_{1+3K} < R_{2+3K} < R_{3+3K}$. This statement says that the rate allocated to UE 1, 4, 7, ... is less than the rate allocated to UEs 2, 5, 8, ... and is less than the rates allocated to UEs 3, 6, 9, ... compared one to one (i.e. corresponding indices compared to each other). Also, we see that for same bit rate requirements, UEs with higher SNR are allocated less resources.

Moreover, as we can see from Fig. 7.62, UEs $3 + 3K$ are higher than UEs $2 + 3K$ which bid higher than UEs $1 + 3K$ due to the fact that they require more resources in view of their applications (5 Mbps vs. 1 Mbps vs 0.25 Mbps). This plot shows the last iteration of the algorithm where the shadow prices are converged. Furthermore, we can observe the coverage area of eNB 20 in Fig. 7.63a, and as we see the coverage is not circular which is a result of REM and elevations in various directions leading to different SNRs. The black addition symbol in the middle of the coverage, shown in yellow, is the eNB 20 coordinates in the 101×101 grid. The side bar shows 0—dark blue—and 1—yellow—for the areas which are, respectively, not covered and are under coverage. The green dots in the region are the UEs scattered all over the area. As we can see, UEs are in the foot print of the eNBs and had good

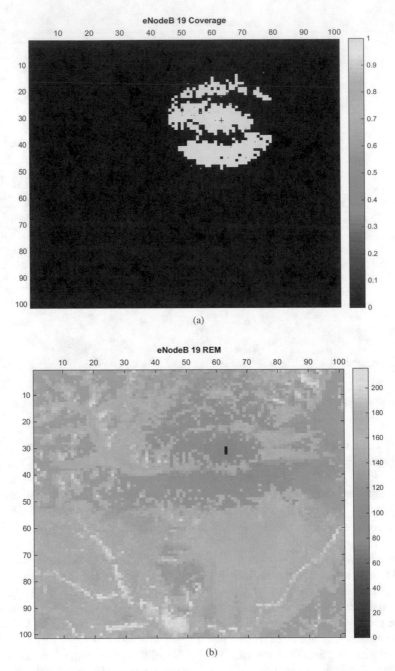

Fig. 7.60 The system contains 23 UEs, each concurrently running a real-time application. (**a**) eNB 19 Coverage. (**b**) eNB 19 REM

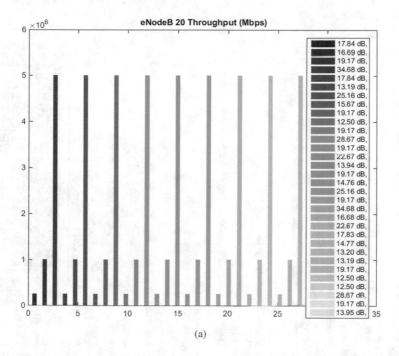

(a)

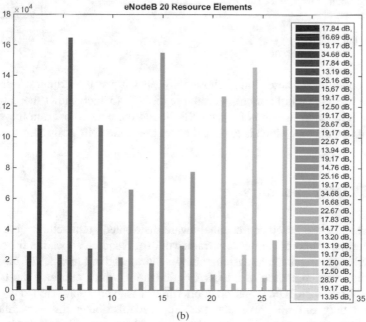

(b)

Fig. 7.61 The system contains 32 UEs, each concurrently running a real-time application. (**a**) Throughput of the UEs for eNB 20. (**b**) Resources Allocated to UEs by eNB 20

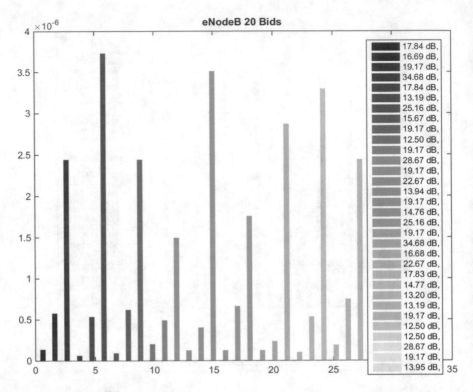

Fig. 7.62 UE Bids pledged to eNB 20

channel conditions. These simply chose eNB 20 as it was the strongest signal they could ever receive. Furthermore, we can see the REM which shows that a blue dot at the eNB location, which is for the eNB 20 and the blue color from the side bar represents a 0 dB loss which is expected as we are at the eNB position.

7.3 Chapter Summary

In this chapter, we developed channel-aware distributed architecture for the QoS-minded utility proportional fairness framework for resource allocation for the cells of a cellular communications system that was introduced in Chap. 3. The distributed architecture was composed of a EURA optimization which allocated the UE rates by the eNB and an IURA optimization which assigned application rates by the UEs. Not only did we prove that the proposed distributed resource allocation architecture's EURA and IURA optimization problems are convex and solved them through the Lagrangian of their dual problem but also we proved the optimality of the rate assignments. We showed that under abundance of resources, the resource allocation assigns more resources to the UEs with bad channel conditions, in order

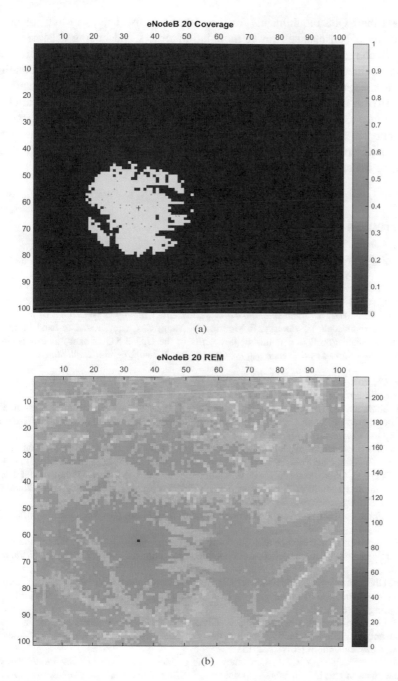

Fig. 7.63 The system contains 32 UEs, each concurrently running a real-time application. (**a**) eNB 20Coverage. (**b**) eNB 20 REM

to meet their QoS requirements for their applications. This is in light of the fact that under a bad channel, lower modulation orders and coding schemes can be used which reduces the spectrum efficiency. On the other hand, when the resources are constrained, more resources are allocated to UEs with good channel conditions so as to meet their bit rate requirements. Ultimately, we performed simulations in MATLAB to show the application of the proposed distributed resource allocation architecture to a cellular communications system. We also performed a large scale simulation through a network planning where the channel conditions was dictated based on the radio environment map and applied the algorithm to assign resources based on the channel conditions.

References

1. J. Reed, N. Tripathi, *Cellular Communications: A Comprehensive and Practical Guide* (Wiley-IEEE, New York, 2014)
2. M. Ghorbanzadeh, E. Visotsky, P. Moorut, W. Yang, C. Clancy, Radar inband and out-of-band interference into LTE macro and small cell uplinks in the 3.5 GHz band, in *Proceedings of the 2015 IEEE Wireless Communications and Networking Conference (WCNC)* (2015)
3. M. Ghorbanzadeh, E. Visotsky, P. Moorut, W. Yang, C. Clancy, Radar in-band interference effects on macrocell LTE uplink deployments in the U.S. 3.5 GHz band, in *Proceedings of the 2015 International Conference on Computing, Networking and Communications (ICNC)* (2015)
4. M. Ghorbanzadeh, E. Visotsky, P. Moorut, W. Yang, C. Clancy, Radar interference into LTE base stations in the 3.5 GHz band, in *Elsevier, Physical Communication* (2016)
5. H. Shajaiah, M. Ghorbanzadeh, A. Abdelhadi, C. Clancy, Application-aware resource allocation based on channel information for cellular networks, in *Proceedings of the 2019 IEEE Wireless Communications and Networking Conference (WCNC)* (2019), pp. 1–6
6. M. Ghorbanzadeh, A. Abdelhadi, C. Clancy, Application-aware resource allocation of hybrid traffic in cellular networks. IEEE Trans. Cogn. Commun. Netw. **3**(2), 226–241 (2017)
7. Y. Chen, M. Ghorbanzadeh, K. Ma, C. Clancy, R. McGwier, A hidden Markov model detection of malicious android applications at runtime, in *Proceedings of the 2014 23rd Wireless and Optical Communication Conference (WOCC)* (2014)
8. M. Richards, J. Scheer, W. Holm, *Principles of Modern Radar* (SciTech Publishing, New York, 2010)
9. M. Ghorbanzadeh, Resource allocation and end-to-end quality of service for cellular communications systems in congested and contested environments, in *Ph.D. Thesis, Virginia Tech* (2015)
10. M. Ghorbanzadeh, A. Abdelhadi, C. Clacy, *Cellular Communications Systems in Congested Environments Resource Allocation and End-to-End Quality of Service Solutions with MATLAB* (Springer, Berlin, 2017)
11. M. Ghorbanzadeh, Y. Chen, K. Ma, C. Clancy, R. McGwier, A neural network approach to category validation of android applications, in *IEEE Conference on Computing, Networking, and Communications (ICNC)* (2013)
12. M. Ghorbanzadeh, Y. Chen, C. Clancy, Fine-grained end-to-end network model via vector quantization and hidden Markov processes, in *IEEE Conference on Communications (ICC)* (2013)
13. M. Ghorbanzadeh, A. Abdelhadi, C. Clancy, A utility proportional fairness radio resource block allocation in cellular networks, in *IEEE International Conference on Computing, Networking and Communications (ICNC)* (2015)

14. M. Ghorbanzadeh, A. Abdelhadi, C. Clancy, A utility proportional fairness bandwidth allocation in radar-coexistent cellular networks, in *Military Communications Conference (MILCOM)* (2014)
15. G.T. V9.0.0, Further advancements for E-UTRA physical layer aspects, in *Measuring of Heterogeneous Wireless and Wired Networks* (2012)
16. F. Sanders, J. Carrol, G. Sanders, R. Sole, *Effects of Radar Interference on LTE Base Station Receiver Performance* (NTIA, U.S. Department of Commerce, New York, 2013)
17. T.U.S.G. Survey, *National Land Cover Data Set* (2005). https://edcftp.cr.usgs.gov/pub/data/landcover/states

Chapter 8
Book Summary

In this chapter, we summarize the material presented in this book.

- We can leverage sigmoidal utility functions to model QoS for real-time applications under a proportional fairness resource allocation framework; and we proved that the optimizations are convex.
- We demonstrated that as a consequence of the optimizations convexity, the allocated rates are optimal.
- We used sigmoidal and logarithmic utilities to model the QoS the real-time and delay-tolerant applications, respectively, and demonstrated that the optimizations for the hybrid traffic originated from sigmoidal and logarithmic application utilities are convex and the solutions are optimal.
- By proving the optimization convexity, we proved that the NP-hard problem of real-time resource allocation, which traditionally was solved via approximating the sigmoidal application utilities to the nearest logarithmic application utilities, can be solved with polynomial complexity.
- We included application usage changes into the resource allocation formulation and demonstrated that higher usage of the application leads to a dynamic resource allocation according to the current focus of the user which leads to miscellaneous application usages.
- We included UE priorities in the resource allocation formulation and demonstrated that higher weights are prioritized in the weight allocations so that the UEs with higher weights are allocated more resources initially.
- We developed a centralized formulation which allocated the rates to the applications in a single stage.
- We proved that the centralized formulation is convex when the traffic is a hybrid of real-time and delay-tolerant traffic whose QoS is modelled by sigmoidal and logarithmic utility functions.
- We obtained the transmission overhead of the centralized resource allocation method.

M. Ghorbanzadeh, A. Abdelhadi, *Practical Channel-Aware Resource Allocation*, https://doi.org/10.1007/978-3-030-73632-3_8

- We analyzed the sensitivity of the centralized method to the variations in the number of UEs in the system and illustrated that the rates retain their optimality when UEs rebid for resources in the face of alterations in the number of UEs in the system.
- We investigated the sensitivity of the centralized method to the changes in the number of UEs in the system and demonstrated that the rates do not retain optimality when the UES do not rebid for resources in the face of the dynamics of the number of UEs in the system.
- We investigated the sensitivity of the centralized resource allocation method to the changes in the applications usage and demonstrated that when the UEs rebid for resources, the rates retain optimality in the face of the application usage changes.
- We investigated the sensitivity of the centralized resource allocation to the variations in the applications usage and demonstrated that when the UEs do not rebid for resources, the rates do not retain optimality in the face of changes in the system.
- We developed a distributed resource allocation method which assigned the rates to the UEs by the BSs and allocated the rates to the applications by the UEs.
- We proved that the distributed resource allocation optimizations are convex and lead to optimal solutions.
- We investigated the sensitivity of the distributed approach to the variations in the number of UEs in the system and demonstrated that when the UEs do not rebid for resources in the face of changes in the number of UEs, the rates do not retain optimality.
- We investigated the sensitivity of the distributed resource allocation method to the variations in the number of UEs and demonstrated that when the UEs rebid for resources in the face of changes in the number of UEs, the rates retain optimality.
- We investigated the sensitivity of the distributed resource allocation method to the variations in the application usage of the UEs and demonstrated that when the UEs do not rebid for resources in the face of changes in the applications usage, the rates do not retain optimality.
- We investigated the sensitivity of the distributed resource allocation method to the changes in the application usage changes of the UEs in the system and demonstrated that when the UEs rebid for resources in the face of changes in the applications usage, the rates retain optimality.
- We investigated the sensitivity of the distributed resource allocation method to the variations in the application usage changes of the UEs in the system and demonstrated that when all applications do not rebid for resources in the face of changes in the applications usage, the rates do not retain optimality.
- We investigated the sensitivity of the distributed resource allocation to the changes in the application usage of the UEs in the system and demonstrated that when the applications rebid for resources in the face of changes in the applications usage, the rates retain optimality.

- We proved that the centralized and the distributed resource allocation methods are mathematically equivalent in that the UE and application rates provided by each method is equal to the ones from the other method.
- We investigated the transmission overhead of the distributed resource allocation and derived mathematical lower bounds for the transmission overhead.
- We proved that the shadow price using the centralized approach converges to an optimal value always.
- We proved that the shadow price using the distributed approach converges to an optimal value if and only if there are more resources available at the BS than the sum of inflection points of all real-time applications utility functions.
- We showed that the shadow price for the distributed approach fluctuates around the optimal value when BS resources do not exceed the addition of the inflection points of the real-time applications' sigmoidal utility functions.
- We provided a technique to stabilize the distributed resource allocation method for all BS available resources by the notion of decay functions.
- We made the resource allocation more pragmatic by considering the LTE radio resource block structure and developed a modified optimization which assigned resource blocks to the UEs.
- We demonstrated that the resource block allocation optimization with hybrid traffic is convex.
- We provided a method to solve the resource block optimization using Lagrangian Relaxation and created a fast mechanism to map the optimal continuous rates to the closest optimal discrete rates.
- We demonstrated that the proposed discrete optimization mapping mechanism decreases the search spaces complexity drastically and yields in a pool of selections for the resource blocks.
- We demonstrated the effect of including channel conditions into the resource allocation modeling methodology.
- The implementation of application-aware channel-aware method may be very cumbersome in terms of run-time. It would make a great project to look at a parallel implementation of the proposed method to augment the pace in order to pave the way for real-world deployment of the method.
- We considered the channel effect for a flat-fading and slow-fading channels.
- We mapped the proposed resource allocation method to the LTE structure by defining protocols for message exchanges required for the resource allocation and finally propelling toward defining standards for the method.
- We looked at the effect of the resource allocation under various channel conditions.

Index

A
Above Ground Level (AGL), 124, 135
Access point (AP), 44
Adaptive modulation and coding (AMC), 94
Aggregate utility function, 50
Application layer QoS, 7
Application utility function, 20
Area prediction mode (APM), 124
Asynchronous Transfer Mode, 7
Automatic Repeat Request (ARQ), 94

B
Bit error rate (BER), 6

C
Call Admission Control (CAC), 7
Cartesian product, 77
Centralized architecture, 37
Centralized resource allocation, 63
Channel quality indicator (CQI), 94
Citizen Radio Broadband Services (CBRS), 135
Continuous rate, 74
Convex, 50, 72
Convex optimization, 38
Cross-layer QoS, 6

D
Decibel (DB), 135
Deep packet inspection (DPI), 4
Delay-tolerant applications, 20
Department of Defense (DOD), 135

Digital elevation model (DEM), 130, 146
Digital signal processing (DSP), 94
Distributed architecture, 48
Dual problem, 51, 107

E
Elastic traffic, 20
Ethernet, 44
Evolved Node-B (eNB), 94, 143
External UE Resource Allocation (EURA), 49

F
Federal Communications Commission (FCC), 136
Fluctuation decay function, 76
Forward Error Correction (FEC), 94
Frank Kelly algorithm, 24

G
Geometric interpretation, 52
Global optimal solution, 49
Great circle, 124

H
Hyper text transfer protocol (HTTP), 44

I
IBM, 41
Inter-cell interference, 82
Internal UE rate allocation (IURA), 49

Internet Protocol (IP), 2
Internet Protocol Television (IPTV), 2
Intracell interference, 82
Intranet, 44
Irregular terrain model (ITM), 124

L
Lagrange multiplier, 18, 52
Lagrangian, 63
Land Use Land Cover (LCLU), 148
Link layer QoS, 4
Log-centralized optimization, 63
Logarithmic utility, 56
Long Term Evolution (LTE), 94

M
Mathematical equivalence, 62
Mobile Network Operator (MNO), 1
Modulation and coding scheme (MCS), 33, 93,
        143
Multiple System Operator (MSO), 1

N
National Elevation Data (NED), 130
Network layer QoS, 5
Noise figure (NF), 94
Nonlinear, 72
Non negativity, 21

O
Open Systems Interconnection (OSI), 6

P
Pareto optimality, 5
Peak-to-average power, 4
Physical Uplink Control Channel (PUCCH),
        97
Point-to-point (P2P), 2
Point-to-point mode (P2PM), 124
Proportional fairness, 24

Q
QoS challenges, 8
QoS Class Identifiers (QCI), 5
Quality of Service (QoS), 2

R
Radio environment map (REM), 123, 145
Radio frequency (RF), 2
Radio resource block (RRB), 71
Radio resource elements (RREs), 74
Rate matching module (RMM), 94
Real-time application, 21
Resource allocation requirements, 33
Resource block allocation, 72, 77
Resource element, 149, 151, 154, 160, 162,
        167, 169, 173, 176, 181, 183, 187, 190,
        194, 199, 206, 213
Resource management (RM), 97
Robust algorithm, 55
Router, 44

S
Server, 41
Shadow price, 52, 108
Sigmoidal utility, 56
Signal-to-noise ratio (SNR), 123
Single-carrier, 71
Sounding reference signals (SRS), 96
Spectrum access system (SAS), 136
Subscriber differentiation, 56

T
Traffic shaping, 45
Transport Control Protocol (TCP), 6
Turbo coding, 94

U
United States Geological Survey (USGS), 130,
        146
Universal Datagram Protocol (UDP), 6
User equipment (UE), 93, 143

V
Virtual Path Indicator (VPI), 7

W
Weighted Fair Queuing (WFQ), 24
World Geological Survey 1984 (WGS84), 126

Printed in the United States
by Baker & Taylor Publisher Services